MW00447207

The Web Designer's Guide to iOS Apps:
Create iPhone,
iPod touch, and iPad Apps
with Web Standards

HTML5, CSS3, and JavaScript

New
Riders

VOICES THAT MATTER™

The Web Designer's Guide to iOS Apps: Create iPhone, iPod touch, and iPad apps with Web Standards (HTML5, CSS3, and JavaScript)

Kristofer Layon

New Riders
1249 Eighth Street
Berkeley, CA 94710
510/524-2178
510/524-2221 (fax)

Find us on the Web at: www.newriders.com
To report errors, please send a note to errata@peachpit.com

New Riders is an imprint of Peachpit, a division of Pearson Education.

Copyright © 2011 by Kristofer Layon

Project Editor: Michael J. Nolan
Development Editor: Jeff Riley/Box Twelve Communications
Technical editors: Zachary Johnson (www.zachstronaut.com), Alexander Voloshyn (www.nimblekit.com)
Production Editor: Myrna Vladic
Copyeditor: Gretchen Dykstra
Proofreader: Doug Adrianson
Indexer: Joy Dean Lee
Cover Designer: Aren Howell Straiger
Interior Designer: Danielle Foster
Compositor: David Van Ness

Notice of Rights

All rights reserved. No part of this book may be reproduced or transmitted in any form by any means, electronic, mechanical, photocopying, recording, or otherwise, without the prior written permission of the publisher. For information on getting permission for reprints and excerpts, contact permissions@peachpit.com.

Notice of Liability

The information in this book is distributed on an "As Is" basis without warranty. While every precaution has been taken in the preparation of the book, neither the author nor Peachpit shall have any liability to any person or entity with respect to any loss or damage caused or alleged to be caused directly or indirectly by the instructions contained in this book or by the computer software and hardware products described in it.

Trademarks

Apple, iPod, iTunes, iPhone, iPad, and Mac are trademarks of Apple, Inc., registered in the United States and other countries. Many of the designations used by manufacturers and sellers to distinguish their products are claimed as trademarks. Where those designations appear in this book, and Peachpit was aware of a trademark claim, the designations appear as requested by the owner of the trademark. All other product names and services identified throughout this book are used in editorial fashion only and for the benefit of such companies with no intention of infringement of the trademark. No such use, or the use of any trade name, is intended to convey endorsement or other affiliation with this book.

ISBN 13:   978-0-321-73298-9
ISBN 10:   0-321-73298-7

9 8 7 6 5 4 3 2 1

Printed and bound in the United States of America

In memory of my father, Roger Layon.
His life taught me to live honorably;
his death taught me to live vigorously.

ACKNOWLEDGMENTS

I'm a runner with a master's degree in interactive design—and the process of writing this book was a lot like marathon training and graduate school. Successfully meeting my goals (all variations of crossing a finishing line) demanded extraordinary levels of planning and commitment.

But equally important was the support of other people. I was really blessed with a lot of support from friends, colleagues, and family—and I thank them all:

The editing, design, and marketing staff at New Riders, Peachpit, and Box Twelve. A special thanks to Michael Nolan, Jeff Riley, and Glenn Bisignani.

Zach Johnson, my technical editor, whose coding experience and critical eye took the book to a much higher level.

Alexander Voloshyn, the creator of NimbleKit, for providing additional technical assistance, several important code samples, and a lot of friendly advice.

Martin Grider and Bill Heyman, who helped me with my first iPhone app and my early efforts to learn Objective-C.

Eric Meyer and Kristina Halvorson, who shared helpful advice and (even more helpful) encouragement.

Mike McGraw at Apple, who helped get me to the 2010 WWDC in San Francisco.

Mark Brancel, my first app client and collaborator. Thanks for your patience and for believing in my work.

Shawn, my friend and legal counsel, whose advice and assistance calmed many a frayed nerve.

Tim, my friend and sailing liberal arts scientist, who taught me how to sail a boat, and who inspires me to see the world differently every time we talk.

Eric, my friend and running coach. The three marathons I ran gave me the discipline and psychological endurance required to finish this book.

My design and communications colleagues in System Academic Administration at the University of Minnesota: Amy, Angie, Gabe, Kate, Kathy, Mike, and Peggy.

My MinneWebCon conference planning colleagues from 2008 to present: Amanda, Dan, Danny, Eric, Gabe, Jesse, Peter, Sara, Simin, and Zach.

My in-laws, Marilyn and Kent, who provide a ton of childcare for us that made this book possible; Marilyn, a writer, also helped edit the first chapter that I wrote, giving me the confidence to submit it to the publisher.

My mother, Sharon, whose skills as a gardener, flower arranger, and stained glass artist elevated my ability to see patterns and beauty, and inspired my own creativity and desire to make things.

My lovely wife and daughters, who gave me the time and space to work on this, and never complained about how tired and unhelpful I must have been during the numerous mornings that followed many late nights of writing and editing: Katie, Sarah, Grace, Emma, and Anne.

ABOUT THE AUTHOR

Kristofer Layon is a designer, educator, and conference director. Kris's first iPhone application, ArtAlphabet, is an early childhood typography flashcard game that went on sale in the App Store in 2009. His consulting company, Aesthete Software, now designs mobile applications for clients in a diverse range of fields including medicine, photography, and education.

He has been a graphic designer since 1993 and a web designer since 1996. Since then Kris has designed sites for engineers, urban planners, city governments, artists, musicians, retailers, the National Park Service, and over 30 higher education clients. In addition to designing websites, he has taught graphic design and typography in the University of Minnesota's College of Design, where he was also an academic advisor. In 2008 Kris helped establish MinneWebCon, a regional conference for web professionals.

Kris holds a Master of Fine Arts degree in interactive design from the University of Minnesota, and a Bachelor of Arts degree in German and pre-architecture from Saint Olaf College. He is a member of AIGA, the HighEdWeb Association, Design Research Society, and Minnesota Interactive Marketing Association. His work has won design awards from the AIGA and the Society of Marketing Professional Services, and his early adoption of web video was featured on apple.com in 1999.

CONTENTS

INTRODUCTION

Here you are, reading a book about designing iOS apps with HTML, CSS, and JavaScript that you can distribute or sell in the iTunes App Store. This must mean that you are a web designer and have some interest in designing native apps for the iPhone, iPod touch, and iPad.

It might also mean that you're ready to take a leap of faith and start reading about something that sounds too good to be true. After all, I had a workshop attendee tell me last summer, "The only reason I signed up for your workshop is because I didn't believe it was possible."

Which, roughly translated into English, means, "I came here thinking you were a liar who wanted to rip me off."

But here's the thing: It *is* possible. And you're now holding the book that I wish I had about two years ago: It doesn't require you to learn how to program in Objective-C, which is really nice for people like me (and perhaps you) who do not think of ourselves as programmers.*

So how does this work, and is this book really a work of nonfiction?

It is indeed. But let's get a few other things straight first.

This book is...

- An introduction to using HTML, CSS, and JavaScript to design native applications for Apple's iOS devices.
- An introduction to using the NimbleKit Objective-C framework, a fabulous collection of library items that allow you to design the Objective-C apps that Apple requires, without having to write any Objective-C yourself.

---

* Of course, HTML, CSS, and JavaScript are all languages that instruct software and hardware to behave in particular ways, so web designers are also programmers. But, still, not really Programmers with a capital P, if you know what I mean.

- A comprehensive guide to visualizing, planning, designing, building, and distributing your iOS apps.

- A manual for designing several types of content-based apps with native iOS interfaces.

- A textbook for anyone teaching iOS app design and content formatting principles to students who want to successfully design their first app before they become grandparents.

- A resource to help app design teams create functional wireframes for sample app navigations and screens.

So that's what this book is. However, it's also important to understand what this book is *not*.

This book is *not*...

- A manual for programming in Objective-C. There are plenty of other books that do this. And remember, NimbleKit already contains all the Objective-C you need—it's written already!

- A step-by-step workbook for designing any app you can think of. There may be apps you can think of that web standards and NimbleKit do not support very well. In that case, you should consider other options, some of which I mention in Chapter 10.

- The complete guide to NimbleKit. NimbleKit is big enough that one reasonably sized book cannot teach you all of it (and yes, I wanted to keep this book reasonably sized so that it wasn't expensive and could be read relatively quickly).

- A collection of the world's best HTML, CSS, and JavaScript code examples. There is usually more than one way to solve a design problem with code. Sometimes I show you more than one way, and other times I just show one. When I choose one, it's either an easier way or just the way I know. If you have another way (and especially a better way), feel free to tell me via this book's website at http://iosapps.tumblr.com. If you submit code that I can test successfully, I will share it with other readers via the website.

- An advocate for Apple's iOS devices or its App Store. Although I am a fan of Apple and its commitment to design and user experience, I didn't write this book from a fanboy's perspective. I'm simply telling the story that I know, and teaching you what I can; both happen to focus on mobile applications for iOS devices.

- An up-to-the-minute reference. Chances are, now that this book is printed, something in it is already out of date. But I'm with you for the long haul: To get updates (and download code samples featured in this book), visit http://iosapps.tumblr.com.

If you're a designer who is familiar with Web Standards, my goal is to open up an exciting new opportunity for you. I hope that reading this book and trying out the examples will lead you to design your own iOS apps, consult with larger design teams on mobile interface and user experience goals, and teach others how to design and format content for use on mobile devices. I also hope that this book is just the beginning. Ideally, it should equip and encourage you to eventually learn much more than what is contained between these covers.

So good luck, and happy reading ... and designing!

1 THE BIG IMPACT OF GOING SMALL

So...why did I write this book?
Aren't there already books
about making iPhone,
iPod touch, and iPad apps?

There are indeed several books about the subject, and they are all very informative. However, I wrote this book for a specific audience.

In short, someone a lot like me.

What I have done is write the book *that I wish I'd had on my bookshelf* about two years ago when I was beginning to investigate how to design iPhone applications. At the time, the only books I could find dealt with programming in Objective-C or explored how to leverage very specific functions and features of the iPhone and iPod touch.

I have nothing against Objective-C programming. I would just rather not do it myself. And while impressed by the features of Apple's mobile devices, I am a designer: What drives my work is not technology itself, but a desire to help people and organizations communicate.

So if you are a designer who enjoys working with people more than wrestling with technology, and solving problems more than experimenting with features, you have found the right book. Because this book approaches the design of iOS applications from a human-centered, need-based approach.

Mobile magic and pocket computers

As iPhones and other smartphones have become ubiquitous, the demand for well-designed mobile content has also increased dramatically. We have all seen some astounding numbers:

- Over 85 million iOS devices sold by mid-2010
- Over 250,000 applications in the iTunes App Store
- 15 billion applications downloaded from iTunes

I have personally experienced the transformative effects of having content available nearly everywhere, whenever needed: while shopping, working out, running, even riding on a chairlift at a ski area. Unless I am at the beach or in the water, my iPhone is usually with me. I can answer questions. I can research something photographed earlier as a reference (I now use the Camera app all the time to take notes), see how far I am from somewhere, check the weather. The list is practically endless.

In fact, I believe the iPhone's name is sort of misleading—it suggests that it's a phone with additional uses. In fact, the device is a networked, pocket-sized computer that you can

- Bring wherever you want
- Use whenever you need it
- Customize by purchasing and installing your own applications

So we think of the iPhone as a phone (**Figure 1.1**) due to its eponymous app, Phone. But Phone is just one of many apps that leverage content, network connectivity, and various hardware and software features to help solve problems or access information when and where you need it.

1.1 Now *this* is a phone! (Whereas Phone is just one of many apps for the iPhone.)

Content—and context—are everything

So why am I focusing on this when it should be pretty obvious already? Because I am making a very important point and helping to shape how you think about designing apps.

To continue this process, consider these two dates:

January 9, 2007
and
May 25, 2010

Do you recognize either date? They are both extremely important for how we should think about iOS apps.

January 9, 2007, is the date on which Apple Computer, Inc., became Apple, Inc. And May 25, 2010, is when Apple, Inc., became the most valuable technology company in the world—three years after it dropped the word "computer" from its corporate name.

The *New York Times* said it best:

"The most important technology product no longer sits on your desk but rather fits in your hand." (May 27, 2010)

Apple saw this coming in early 2007 when they changed their name, perhaps because the iPhone was on the horizon for later that summer.

But it was not the iPhone alone that made Apple the biggest tech company by the spring of 2010. The process started in 2001 when the iPod was introduced and continued in 2003 when iTunes was launched. And note that although they did not stop making full-size computers at the time, they started making some that were much smaller. And this helped integrate computer technology into our lives much more than desktop and laptop computers ever did.

With the launch of the tiny new iPod computer platform, Apple took a leap that was much larger than the one they first took from the Apple II to the Mac. In that first stage of evolution, Apple popularized the graphic user interface (GUI), the visual desktop metaphor, and the mouse input device to create a whole new world that now pervades all personal computing.

Today most people interact with and work on personal computers without needing to speak their languages (that is, actually program them to do all the work).

The most incredible thing about the iPod is that it's pared down from a Mac quite drastically. Much more. Apple not only took away the mouse, they took away the desktop. They made the screen incredibly small, and made it impossible to create any content directly on the device (**Figure 1.2**). So Apple took a computer, removed a lot of its functionality, made it as simple and small as possible, and made it completely unproductive. And, thus, utterly unbusinesslike—the exact opposite approach to their personal computing strategy of the 1990s.

And the result of this drastic reduction in power, size, and capability?

Sales went through the roof and have made Apple incredibly successful.

1.2 The original iPod: Shrinking and simplifying the computer even more than the Mac did.

This transformation is profound, because we don't even think of an iPod as a tiny computer. Instead we think of it as a portable, practical, and easy-to-use device for listening to music, news, information, and audio books, and seeing photos or even watching movies and TV shows.

This isn't business content. This is life content.

Interestingly, Apple didn't break much new technological ground with the introduction of its iOS devices. Rather, when introducing the iPhone in 2007, they simply added back a few of the key features they had taken away when they made the enormous leap from the full-size computer to the tiny iPod platform. They restored the Internet connectivity that we're used to at our desktops and laptops, and reintroduced the ability to enter information via a (screen-based) keyboard after previously limiting users with the iPod's click-wheel.

And oh yeah...the iPhone got the Phone application!

More importantly, iOS devices push the life content concept much further. Now the news can be breaking, the music can be live, and the information can be our kid's soccer schedule. Or a restaurant's address, shown on a map, with directions from our current location.

All of this means that designing for these devices needs to begin by focusing on life content, and centering on the lives of human beings and the problems that we need solved on a daily basis.

To design for these circumstances, we need to keep it real.

Mobile applications ≠ desktop applications

But we also need to keep it simple.

Thinking about mobile content needs from a life content perspective allows you to focus on the proper context of your design work, which is how and when people seek information. Next we need to focus on how people actually use mobile devices.

Many books and presentations about app development focus on leveraging specific features (learn how to make the device vibrate!). And to be sure, for a large software team working on a complex application, it might make sense to focus on specific technical features and behaviors. But this book is written from a human-centered perspective, which puts people and their content first.

When people use mobile content, their paramount concern is probably not whether someone has employed particular features of the device. But if designers neglect certain features or implement them badly, people will undoubtedly notice. Staying focused on content and on people's needs should lead us to adopt the right behaviors and features for the right reasons. This makes much more sense to me than learning how to implement a feature first, then trying to think of a way to build an app around the behavior.

Designing an app for a mobile device is very different than designing a website that will be viewed on a computer with a full-size screen. We tend to use desktop—and even laptop—computers more often in work or educational situations. (Obviously the Internet has changed that dramatically, but bear with me a moment.) This means that the places where we tend to use computers are in offices, classrooms, or at home. Sure, a laptop can be brought many places and some of us have them with us nearly wherever we go, but most people use them in one primary location.

Compare this to using an iPhone, iPod touch, or iPad. Depending on your device and which model you have, your connectivity varies, and yet, given the pervasiveness of wireless networks, having one of these mobile computers in your pocket (more metaphorical for an iPad owner—unless you have *really* large pockets!) means that you are using it on the go. And this means the context is often different, and the reason for using it is probably *entirely different*, than using a computer. Double-checking a recipe for its list of ingredients is not a business-related task unless you are a chef or caterer: You take your iPhone or iPod touch out at the grocery store because you are human, you are hungry, and you need to solve this particular problem on the fly.

And it's not a technical problem.

The most important thing to remember is this: Designing content for an Apple iOS device is different from designing content for a browser on a larger screen. And it's about more than just context. On a full-size computer, the browser transforms a bit to contain our content, but we really don't think of it that way. The browser shows us a website, and it is still on our computer—we don't really perceive the browser itself changing.

The magic is transformational

So here's another key difference with an iOS device: Our content delivery, particularly in a native app designed very specifically to support a particular communications need or end-use, *becomes* the thing we've designed.

If you're not convinced, compare the ratio of the screen's surface area to other hardware or surface details on the two devices.

The iPhone, iPod touch, and iPad are very similar in one respect: They are mostly screens. All three are about 95 percent screen from a frontal perspective, whereas a laptop is a bit less than 50 percent screen because of the keyboard and a lot of additional surface area (**Figure 1.3**). That's a substantial difference. Even when an engaging website is showing on a laptop screen, it's still visibly *on* the laptop, isn't it? The keyboard, touch pad, wrist area, and frame around the screen do not disappear. They still impact the experience and keep us more distant from the content.

1.3 Even laptops are mostly keyboard, touch pad, and frame. The screen is less than 50 percent of the overall surface.

But as soon as you run an app on an iOS device, the entire device seems to transform because of its multitouch, screen-dominated design. This is more obvious in some apps than in others, but consider how the iPhone appears to become a phone when the Phone app is running. The hardware nearly disappears: Suddenly, we have a glowing phone keypad with built-in list of contacts. The same is true with Maps: It's not really a map inside a device; the app helps the device *become a map*.

Understanding this transformational effect is critical to our approach in designing for these devices. We need to be extra careful about how we design a user interface (UI). We need to learn how to respect the native Apple iOS controls, and when to design custom UI elements that directly support the communications needs of the app we're designing. And, in some cases, the UI elements extend beyond mere functionality, as branding might also be a factor.

It is also essential to understand just *how* important this design thinking is in this situation. Because I would argue that missing the mark in either native Apple UI controls or content- or branding-specific UI detailing is not the same as handing in a paper for a class and having it be just good instead of excellent.

Unfortunately, the illusion of the device turning into something else is far more picky than that: Missing the mark by any measurable distance doesn't just result in a very good instead of an excellent app. It can too easily result in an extremely unconvincing, or even annoying, app.

"Well, wow, now this is starting to sound confusing," you might be thinking. A well-designed app should be immersively designed to fit seamlessly into the user's life no matter where they are, feature native Apple iOS user interface details wherever appropriate, *and* include content- or branding-specific UI details when appropriate? What's the magic formula for that, anyway? And how do we focus on both content and user interface enough to pull this off successfully?

The answer is, of course, that there isn't one formula. But fortunately web designers already have experience in responding to client and customer needs, and are familiar with designing using corporate style guidelines. To apply this valuable web design experience to iOS app design, you need to become familiar with native user interface standards, details, and recommendations; define your project requirements (whether for your own project or that of a client or employer); and determine the best way to design it for iOS devices.

This book will continue within this paradigm, and show you how to leverage your design experience and web skills into designing iOS apps, and introduce you to some techniques that allow you to craft these projects without writing your own Objective-C code.

Design starts with people and ends with code

"But wait," you're thinking, "I thought that all apps have to be written in Objective-C?"

That is true. However, this does not mean you need to write the Objective-C yourself!

In fact, what if someone else has already written it for you?

Consider JavaScript frameworks for designing content to be displayed in web browsers. Two of my favorite examples, jQuery and Yahoo! User Interface (YUI), assist web designers with employing JavaScript-powered behaviors by doing most of the heavy code lifting for us (which begs the question: How much does code weigh?). The code to achieve some very nifty behaviors has been pre-written for us into modules, and we just hook into these and use them without having to write everything from scratch. In fact, don't forget the other important aspect of using a framework: The code has been thoroughly tested (and continues to be tested and updated on an ongoing basis) so we don't have to debug our own code, either!

Similarly, there are iOS design frameworks that do the same thing for iPhone app design. The framework I've used the most is called NimbleKit (**Figure 1.4**). This awesome tool has a bunch of Objective-C—already prewritten for us—that performs native iOS functions and behaviors, and has been developed to be called into action with designs that use HTML, CSS, and JavaScript. When you think about it, it's very similar to using these same languages to make a web browser display particular content and behave in a certain way. We're simply using a different code framework in this case, and designing for a particular operating system (iOS) and distribution network (iTunes).

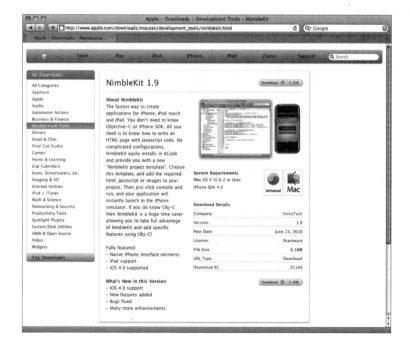

1.4 NimbleKit is a featured development tool on Apple's website.

That is what the subsequent chapters are all about. Learning more about the important characteristics of iOS interfaces and behaviors, seeing how NimbleKit bridges the gap between familiar web design languages and new devices (and their native development languages). And, after learning a few examples of designing content-based iOS apps, we'll talk about how to submit them to Apple for approval and how to distribute or sell them in iTunes.

By following Apple's lead from 2001 to the present, we can see that content reigns supreme, and that small, simple devices with intuitive and consistent interfaces and behaviors bring our most useful content—life content—to people where and when they need it. All we need to do is focus on people and what they really need, summon our web design skills, learn some new tips and tricks, and hitch our wagon to Apple's very successful train.

We're just getting started!

Summary

What you've learned in this chapter:

- Don't let the name "iPhone" fool you. It's a pocket computer with a Phone app. This means it can do amazing things, and we can help shape some of these things!

- The entire iOS ecosystem is an enormous change in strategy for Apple in that it focuses on lifestyle, ubiquitous information, and content. This has created an enormous opportunity for people who design digital things.

- iOS use is immersive and more fully integrated with the rest of our lives, not just our work. How can we design apps that fit into this context?

- Native apps are programmed in Objective-C, but that doesn't mean that designers need to learn Objective-C. They can work in teams with programmers, or use a code framework to bridge the gap. This book is primarily about the latter.

2 ESTABLISHING YOUR APP DESIGN STUDIO

As we start getting into the technical and production-oriented aspects of iOS application design, I want this book to do you a favor—a *really big* favor: I want it to answer more questions than it raises.

This may be more of an aspiration than a truly achievable goal, but it is still a sincere goal and one born out of my own app design experience. For one thing, many of the processes require more steps (and thus, more decisions to make) along the way than I first anticipated. And as I encountered all of these steps, I rarely encountered enough documentation to get me through the entire process easily and safely.

So I hereby decree that the rest of this book might get a little tedious at times, but will do so in order to give you enough detail to master some very fundamental aspects of iOS app planning, design, and production. And while it may not teach you to design *any* app that you want, what it does teach you it will teach you very well.

In order to design iOS applications, you need to establish what I call your app design studio. This studio will be the place in which you design, test, and package your applications for submission to Apple.

The foundation of your studio will be an Intel-based Mac with the Snow Leopard version of Mac OS X (operating system). And atop that foundation will be Xcode, the integrated development environment (IDE) created by Apple so that you, in turn, can design software for Apple. But while Apple designs some of its software for both Mac and Windows computers (such as iTunes and Safari) and millions of iOS device owners sync their devices to Windows computers daily, Xcode doesn't run on the Windows OS.

Once you have a Mac with the right processor and operating system, I could theoretically tell you to just download and install Xcode, and you would be ready to design iOS apps. But getting the free Xcode is not just a simple, one-step process, so let's walk through how to do it in greater detail. We'll be thorough, yet as quick as possible. After all, we've got apps to design!

Getting an Apple Developer ID

To install Xcode, you need to download the iOS SDK (formerly the iPhone SDK; the name changed in July 2010). But before you search for that, you need to get an Apple Developer ID.

The Apple Developer ID is your online identity for the entire app design, review, distribution, updating, and (if you're selling your apps) compensation process. And, I suspect, because Apple would like you to eventually participate in this entire process, they require you to register for the Apple Developer ID before you can even download the iOS SDK. But fear not: The Apple Developer ID itself is free, as is downloading the SDK.

To start the ID process (**Figure 2.1**), visit **http://developer.apple.com/programs/register**.

2.1 The starting point for obtaining your Apple Developer ID.

The second screen in the process (**Figure 2.2**) asks whether you have an existing Apple ID, and explains that you can, if you wish, use an Apple ID that you already have for iTunes or for purchasing goods online from the Apple Store. Yet Apple ominously notes under these choices that it is probably better to create a separate Developer ID to "avoid accounting and reporting issues."

2.2 Unless you are already an Apple Developer, don't use an existing Apple ID. It could make accounting and reporting problematic down the road.

One has to wonder: Why are we even given an option here?

If you're like me, you probably dislike accounting or reporting problems, so I have a strong suggestion: Just create another ID. This keeps your Apple *customer* role completely separate from your Apple *Developer* role. It also allows your Apple Developer ID to be associated with a corporate entity with its own tax identification number, which may offer some tax advantages to you down the road.

After selecting the new Developer ID option, you'll be forwarded to personal and professional profile pages. This is another area that appears to require careful answers, but don't be particularly concerned. For example, there are questions about your primary app market, the categories of apps you plan to design, and whether you design for other mobile platforms. I have yet to hear any stories about Apple not approving an app because it doesn't fit a category that was selected in this profile. Similarly, there are probably hundreds of Apple Developers who also design apps for other mobile platforms. So I don't think you'll be penalized for being honest if you're also, say, an Android app designer. (On the other hand, why is Apple asking? Just chalk it up to one of the many unexplained mysteries that surround Apple.)

Apple Developer Agreement

After completing your personal and professional profiles, you are forwarded to another slightly intimidating screen: the Developer Agreement page. Again, don't overthink this. On the other hand, this is more than just your standard EULA that you toss aside after buying new software. It outlines a legal agreement that you are making with Apple, as a developer of software for their devices. So what does *that* entail?

Now I'm not a lawyer, so don't read on thinking that you'll get a bunch of legal advice about the merits of Apple's Developer Agreement, but here are some things to be aware of:

- Apple Developer ID and password: Keep these confidential and don't share them with anyone. Unless you want to share them (and I'm not kidding here) with your child aged 13 to 17 who wants to design software for Apple under your ID. Isn't it reassuring to know that you can start a family business with only one Developer ID?

- Developer benefits: No, you don't get health insurance or paid vacation by registering to be an Apple Developer. But there is one particularly nifty benefit: being eligible to attend Apple's annual Worldwide Developer Conference (WWDC) in San Francisco. This is the legendary event where Steve Jobs announces big news (like the first iPhone in June 2007, and the iPhone 4 in June 2010). The first day of announcements and updates is followed by four days of technical sessions and labs about all aspects of Apple's iOS and Mac technologies. It is attended by Apple Developers from all over the world and is a great educational and networking opportunity. But don't think that this is a free benefit—the 2010 registration fee was $1,600.

- Restrictions and confidentiality: Once you agree to become an Apple Developer, you're also agreeing to not share the juicy details about how Apple's devices and software work, or exciting things that you learn at the WWDC, or any advance notices about new Apple products. Essentially, you are signing a nondisclosure agreement. (This can be a fun thing to act coy about with your friends: "Well, I do have some accurate information about that Apple rumor, but unfortunately I'm not at liberty to tell you because I'm under NDA.")

THE NARROW LINE I HAVE CAREFULLY TROD

Because I'm party to the Apple Developer Agreement, I have to be careful about the information I include in this book. So, for example, rather than go into great detail about Apple's human interface guidelines (which are documented extensively at developer.apple.com), I'll convey the spirit of those guidelines and encourage you to download the full documents and add them to your reference library. Basically, I'll limit my instruction to the big picture: giving you an overview of how to register as an Apple Developer, design apps, get them approved by Apple, and distribute them via iTunes. Given that Apple currently keeps 30 percent of gross app income, the more people I can orient to these processes, the happier Apple will be! On the other hand, an Apple Developer is not an official agent or representative of Apple, so I'll draw the line at summarizing or restating any of the more detailed information that is available only to developers on Apple's private developer websites. Therefore, the only representations of protected Apple websites (which require authentication with a Developer ID) included in this book are

1. Screenshots of the iOS SDK download process
2. Screenshots of my own iTunes Connect account

- Apple trademarks and logos: As an Apple Developer, you are granted limited use of Apple marks. For example, you can include the "Available on the App Store" badge on your website, as well as some very nice photos of iOS devices that Apple provides to you. You are also allowed to mention Apple product names in your site's content. In either case, be sure to follow their usage guidelines (highlighted in chapter 11) and include any required disclaimers.

The agreement is relatively benign and has some benefits as well as constraints. I suggest reading it carefully and agreeing to it, but again, I am *not* a lawyer (I do not even play one on TV).

Verification and celebration

After you agree to the Developer Agreement, you'll need to complete a quick verification process in which a code is emailed to you. Once you receive and enter this verification code in the proper field and submit it, you are presented with the congratulatory acceptance screen as Edward Elgar's *Pomp and Circumstance* plays in the background (**Figure 2.3**).

2.3 Congratulations, you're an Apple Developer! (Minus the applause and *Pomp and Circumstance* playing in the background.)

You've made it—you have a Developer ID, and are on your way to downloading what you've been patiently anticipating: the iOS SDK!

Downloading and installing the iOS SDK

Whew! Fortunately the rest of this process is quick and easy. Well, mostly easy. And, well, not so quick—that depends.

The fact of the matter is, the iOS SDK is one of the largest software downloads I have *ever* downloaded to my computer. Ever. At more than a whopping 2 gigabytes, I have run out of creative ways to warn attendees about downloading this in advance of coming to my app design workshops. I mean, *you feel yourself growing old* as it downloads. So if you try this at home and you're on cable or slower, start the process and take your dog for a long walk while it downloads. But please don't sit and watch it download. And if you do, don't say I didn't warn you.

The good news is, the download is really worth it. You get some dandy tools that do some incredible things, and they are all free.

Yes, this is the one free part of being an Apple Developer. You get their developer tools gratis. And it's a nice set of tools that includes two essential programs: Xcode (for designing and packaging your app code) and Simulator (for testing your app along the way).

NOTE Other things in the SDK

Another useful tool in the download is Dashcode, which has a more graphic and object-oriented approach to designing web apps that live on servers (rather than being downloaded and installed on devices). While this book does not cover web app design, many of the native app design skills you'll learn here can be applied to designing web apps, which can then be viewed on any mobile device with a standards-based web browser. There are several other tools in the SDK, too, such as Interface Builder and a variety of diagnostic tools. These are not necessary for the apps and methods covered in this book, but might be of interest if you decide to try your hand at writing your own Objective-C.

To start your adventurous download (**Figure 2.4**), visit **http://developer.apple.com/devcenter/ios/**

2.4 The iOS Dev Center, your resource for iOS development.

Under the page heading, there is a View menu where there are often (but not always) two tabs: one for the current released version of the SDK and one for the beta version of the next release. I encourage you to download the released version. The betas are really only useful for more advanced application designers who are staying ahead of the curve; the betas help them design apps that take advantage of new device or software features before they actually hit the marketplace.

NOTE **Directory location of the SDK**

Note that Xcode and its related iOS SDK apps are not installed in your regular Applications directory. If you went with the default location, these applications were installed on your drive's Developer directory, just one folder in from the root.

Assuming you are just downloading the SDK to design for the current release, proceed with that process. Once the download is complete, you should have a large .dmg file in your Downloads directory. At the time of this writing, the iOS SDK 4.1 download was 2.17 gigabytes. Double-click it, follow the prompts and directions, and when you reach the end you'll have the first part of your design studio in place!

Now it's time to download and install NimbleKit, your Objective-C framework.

Downloading and installing NimbleKit

Aren't you glad to read that the last step is super-easy and requires no elaboration? I bet you are!

1. To download NimbleKit (**Figure 2.5**), visit **http://www.nimblekit.com/**.

2.5 The NimbleKit website.

2. Click the Download NimbleKit link. This time you'll have a much smaller file, nimblekit.dmg, to save to your Downloads directory.

3. Open the file, follow the prompts and directions, and the NimbleKit framework will be installed.

NOTE The NimbleKit license
You probably noticed the Buy online link in the right margin of NimbleKit's home page. My advice: Ignore that for now. You can download NimbleKit for free, design apps with it, and test them in Simulator, all without paying the $99 fee. This is very nice, because it allows you to learn a lot before making a commitment. On the other hand, NimbleKit's fee is one-time only (unlike Apple's annual $99 Developer fee).

Summary

What you've learned in this chapter:

- Every step of a new design project, and process, has consequences, including obtaining and setting up your tools! So think it through.

- Get a unique Apple Developer ID so you keep your iTunes purchases separate from your interactions with iTunes and iTunes Connect as a developer.

- Read the Apple Developer Agreement carefully and make sure you are comfortable with what it says. Like any agreement, it creates opportunities as well as defines constraints.

- Downloading and installing the iOS SDK provides you with the foundational applications—Xcode and Simulator—you'll need for the rest of this book.

- Downloading and installing the NimbleKit Objective-C framework enhances Xcode with a library of prewritten Objective-C resources, allowing you to focus on app planning, interface design, and content formatting instead of learning a new programming language.

Congratulations, your design studio for the upcoming exercises is now ready for action! Next we will explore Xcode and how to begin an app project.

3 FUNDAMENTALS OF THE IOS SDK

Chefs have kitchens.

Stylists have salons.

Physicians have clinics.

Mechanics have garages.

And Apple mobile app designers have the iOS Software Development Kit (SDK).

Like the rest of these workspaces, you won't necessarily use all the available features all the time. In fact, you may never use certain features. And if you're like me, you'll probably never try some of the things that are available in Xcode.

But we can't design apps without Xcode and Simulator, so this chapter explores some basics about how to use them as you work your way through the rest of the book. And, you'll begin to appreciate just how well you've tricked them out with the addition of NimbleKit!

Xcode, the integrated development environment (IDE) that Apple has created for aspiring app designers like us, is an amazing computer program. And I tend to put it in the same category as Adobe Photoshop: It has so many settings, menus, submenus, and features that I know I'll never master it. It's just too big and, frankly, I have yet to find a reason to try.

And after Objective-C itself, using Xcode is one of the more challenging aspects of exploring iOS app design. Fortunately, adding a code framework such as NimbleKit makes both Objective-C and Xcode much easier to put into action. So after learning some basics, we'll dive directly into app design.

This chapter provides an overview of Xcode from the beginning to the end of the app design process, so you might find yourself coming back to this chapter repeatedly to remind yourself of Xcode-specific issues throughout your app design journey.

Starting a new Xcode project

When you start a new project in Xcode by clicking on the File menu and choosing Project (File > New Project), you're presented with a number of choices (**Figure 3.1**). Several of them relate to the types of navigations and views that are available, while others relate more to the app's function or mode of operation. For us, the choice is simple: We're going to create a NimbleKit application.

3.1 Starting a new NimbleKit project in Xcode.

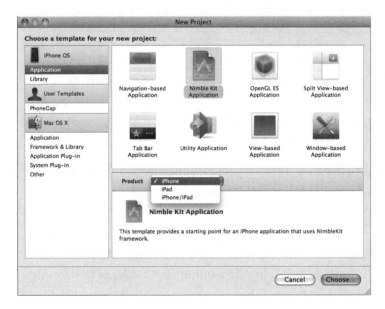

Under the app type choices, you also see a Product drop-down menu with the options iPhone, iPad, and iPhone/iPad. These choices might initially seem a bit confusing — after all, if you design an iPhone app won't it work just fine on an iPad, too? Yes, it will. But there are still some things to consider with the three main options being presented:

- iPhone project: If the above reasoning is good enough for you, and you're not designing a separate iPad version of your app, then selecting the iPhone project is sufficient. And, it covers both iPhone 4 and pre-iPhone 4 models.

- iPad project: If your app is designed explicitly for the iPad in one way or another (via interface, content, or both), this is the option for you. This is also the route to take if you have an iPhone version of your app but you want to, for example, release an iPad version separately as a premium version.

- iPhone/iPad: This is also considered a "universal app," and contains screen layouts and graphics scaled for all iOS devices. That way a customer need only purchase a single version of the app and that version runs natively on the iPhone as well as the iPad. This might result in a happier customer if he owns more than one Apple iOS device and it's an app he would want on all of them. Or, it could disappoint you, because you just sold two versions of your app for one price.

NOTE

For working through this chapter's examples, create an iPhone project.

I'm not going to steer you in one direction or another with app types, as it truly depends on how you want to reach your audience, the value you want to deliver, what you want to earn, and how well your app runs on either platform. I don't believe there is just one best answer—you need to trust your gut but, more importantly, you need to design intentionally for one device or the other, or both, as your content and other purposes best suggest.

After you click Choose, you're prompted to name your project. At this point, don't fret too much about the final name of your app, as it has nothing to do with the name of the project file. You can always rename your project later (see "Naming a Project and Creating the App Bundle" later in this chapter).

Groups & Files, Detail View, and Editor View panes

Once you've created your new project, you're sent to the main Xcode view
(**Figure 3.2**). The first thing you'll notice is that it looks like a souped-up
Mac OS X Finder window. In many respects, that's exactly what it is and—
especially in the case of using an Objective-C code framework—this is also
how we'll use Xcode much of the time.

3.2 The three panes of
an Xcode project window.

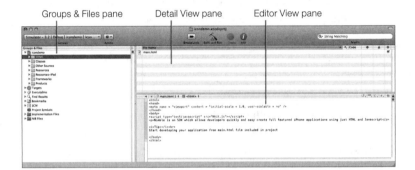

Groups & Files pane   Detail View pane   Editor View pane

Under Xcode's toolbar are three primary areas, or panes, of information
that function similarly to the panes of a Finder window. On the left of
the window is the Groups & Files pane, with the .xcodeproj file at the top
(named icondemo in Figure 3.2). Before you click around, this file is highlighted by default and you see all of the project's other file components in
the Detail View to the right.

The third pane of an Xcode project window is the Editor View. As you
might expect, the biggest difference between Xcode and a Finder window is
the Editor View; this is where you do most of your app work within Xcode.
In the Groups & Files pane, click on a right-pointing triangle to the left
of a folder to reveal its file contents, and then highlight a file. Its contents
appear in the Editor View pane. This code becomes the new default view of
this pane; even when you go back to highlighting folders, the Editor View
continues to display the contents of the last file you selected.

In fact, the Editor View always shows the last file you worked on, even if
you save and exit the project and reopen it.

The NimbleKit file structure and contents

Before we go any further, now is a good time to mention what is in the folders of a new NimbleKit project. As I provide more details, please keep your web designer hat on, because there are similarities to designing sites here that should be familiar to you.

To begin, the top folder is entitled HTML and contains exactly that: your starting content in the form of a single file named main.html. Highlight the file and you'll see this in the Editor View:

```
<html>
<head>
<meta name = "viewport" content = "initial-scale = 1.0,
  user-scalable = no" />
</head>

<body>
<script type="text/javascript" src="NKit.js"></script>
<p>Nimble is an SDK which allows developers to quickly
  and easy create full featured iPhone applications
  using just HTML and Javascript</p>
<i>Tip:</i><br />
Start developing your application from main.html file
  included in project

</body>
</html>
```

You can also stash additional files in this HTML folder and treat it like the root directory of a server for a website. So CSS and image files can all live here together.

Before we proceed to editing anything or adding files, it's important to understand that the remaining folders and files are basically off-limits. Don't become curious and edit anything in the Classes, Other Sources, Resources, or Frameworks folders. Do! Not! Do! It! (**Figure 3.3** on the next page).

NOTE

Web designers typically organize their markup by placing JavaScript in the head of an HTML file. The main.html file in the NimbleKit download places it in the beginning of the body instead. Not how I typically do it, so I wanted to mention this. (Still, it doesn't matter where it goes as long as it's before the `</body>` tag.)

3.3 When you see files like the ones shown here, with .h, .m, .xib, and .framework file extensions, do not edit them. Editing them will disable the NimbleKit framework!

Why shouldn't you edit these files? NimbleKit comes with a veritable treasure chest full of great code in these folders, just like code frameworks for websites. And all of this richness is in these folders. So just as you would never download a JavaScript framework like jQuery or a CSS framework like Blueprint, and then start messing around with the files, you shouldn't do anything to the files outside the HTML folder until explicitly instructed to do so. (There are a few isolated reasons to make changes in some of the files there, and I will carefully explain why at the end of this chapter.)

Naming a project and creating the app bundle

I should probably not admit to this, but until recently, I foolishly spent a lot of time manually renaming a variety of files in my Xcode projects. The app name can inform the name of many other things including some of its key resource files, so changing the name manually requires you to change it in several different files and locations. But then I learned that there is a much, much easier way to do this.

Aren't you glad I figured this out in time to tell you?

So here is the secret: Project > Rename.

I know—it could not have been labeled any more clearly, right? Well, fret not—I am feeling plenty sheepish.

When you select Rename, you're prompted with a small window that allows you to select up to five items to include in the renaming process (**Figure 3.4**).

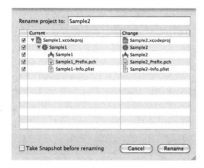

3.4 Renaming an app project.

These are five of the items that comprise a project's *app bundle* (more on this later in this chapter):

- Project file (name.xcodeproj)
- Target (name)
- App name (name)
- Precompiled header (name_Prefix.pch)
- Information property list file (name-Info.plist)

For consistency's sake, I suggest checking all of them so your name is propagated across all of these items.

There appear to be no significant limitations to project or app names. For example, Apple limits app names to 255 characters. I'm not sure why you would ever want an app name that long, and neither is Apple because they recommend that the name not exceed 35 characters. At that length, the entire name shows up in the App Store app on iOS devices. **Figure 3.5** shows a filename that is too long, indicated by the ellipsis.

3.5 This is what you get when your app name for the home screen is too long—a truncated version with an ellipsis.

But there's a more important limitation to the name that you probably want to consider: The name that can appear under the device icon is much shorter than 35 characters. And to make it somewhat complicated, there is no set maximum because Helvetica is not a monospaced font, so the more narrow characters in your app name (i, l, j), the longer your app name can be; the wider the characters (O, D, Q), the shorter. Practically speaking, an app name that fits under its screen icon is 10–12 characters in length.

NOTE Close also counts in screen names and iTunes names

The app's screen name doesn't need to be identical to its name in iTunes. For example, you might use Wonder App as a screen name but Wonder App iPad Edition for the full iTunes name. Apple prefers that both names are the same (or very similar to each other), and Apple has other requirements, for example, the name cannot be trademarked by others. For all of these details, check the iTunes Connect Developer Guide.

Making your app icon

Another key part of your app's identity is its icon. A new NimbleKit-based Xcode project does not come with a default app icon, so you will have to make one.

Or, several. How about up to seven for a universal app?

Until April 2010, designing an app icon was a simpler process. You just made three sizes: a small one for the device's home page, a smaller one for Spotlight search results, and a much larger one for the iTunes App Store.

But with the advent of the iPad and then the iPhone 4, things got really interesting with the app icon because each of these devices has a resolution that is different from the iPod touch and earlier iPhones. So now each screen size needs its own home screen and Spotlight search icons if you want them optimized for their native resolutions.

To be perfectly honest, I am not convinced that the Spotlight search icon is particularly important. If your icon is designed to be viewed at 57 × 57 pixels, there's a good chance that the results you get by resizing to 29 × 29 pixels will not be noticeably better than what the device does when it scales it for you. And do you really want to design an icon that is optimized for 29 × 29 pixels? Unlikely — it's quite a tiny size for trying to optimize any design.

But if you are a pixel perfectionist and extremely picky about graphics being scaled properly, you will want to resize (or redesign) your app icon as needed for each use scenario. **Table 3.1** shows the specifications for the icon files that are bundled with the app.

TABLE 3.1 Size (in pixels) and name specifications for iOS app icons

APP ICON	IPHONE / IPOD TOUCH	IPHONE 4	IPAD
Home screen	57 × 57 (Icon.png)	114 × 114 (Icon@2x.png)	72 × 72 (Icon-72.png)
Spotlight	29 × 29 (Icon-Small.png)	58 × 58 (Icon-Small@2x.png)	50 × 50 (Icon-Small-50.png)

For the iTunes App Store, the icon is 512 × 512 pixels (but the name is not strictly specified by Apple like those that are bundled with the app). And regarding format specifications, save all your icons in PNG format with a bit depth of 24 bits and no transparency.

Though your icons will probably not vary like these examples (see "Sample icon files: Download them from the web!" for why they vary here), this is an otherwise complete set of app icons for a universal app. Note that the iPhone 4 icons are twice the size of the regular iPhone app icons, even larger than the iPad icons (**Figure 3.6**).

iPhone app icon   iPhone Spotlight app icon   iPad app icon   iPad Spotlight app icon

3.6 The full range of home screen icons and Spotlight search icons for the iPhone, iPhone 4, and iPad. The first six are part of the app bundle; you submit the 512 × 512 pixel icon separately to iTunes for display in the App Store.

iPhone 4 app icon   iPhone 4 Spotlight app icon   App Store icon

---

NOTE   **Sample icon files: Download them from the web!**
I like sample files, and I bet you do, too. So I've created a sample set of icon files for you to download. Just visit http://iosapps.tumblr.com and download the app-icons. zip file. When you open it, you'll have a sample app called icondemo and the set of files shown in Figure 3.5. I've made each one a different color so you can see how the Simulator loads different files depending on whether you're testing as an iPhone, iPhone 4, or iPad. And I made three different grayscale versions so you can see how they come up in a search.

There are two more steps to successfully getting your app graphics into your app:

1. Add them to the project.

2. Add their file names to the information property list, or name–Info.plist, file.

Finally, note that Chapter 12, "Using iTunes Connect and the App Store," provides some additional guidelines and suggestions for how to graphically design your app icons.

Adding files to your project

Adding files to a project is another example of where Xcode is a lot like the Mac OS Finder. While you can't use Finder to drag your new icon graphics from where you first saved them to your project directory (and have them show up in the project), you can drag them from a Finder window to your Groups & Files pane in Xcode. You can also add them via the Project > Add to Project menu.

In either case, I suggest that you store your app icons in the Resource directory of your app. It's important to note that the folders of an app project that you see in the Groups & Files pane are for your benefit, not Xcode's. It's a flat file structure, so where you keep your files is usually unimportant (though there are a few exceptions). In this case, I simply suggest that you keep the icons in Resources so they don't clutter your HTML directory. Plus, the HTML directory is for app content, so the app icons don't really belong there anyway.

When you move the icons into the Resources folder using one of the aforementioned methods, be sure to check the box labeled Copy Items Into Destination Group's Folder (If Needed) before clicking Add (and you can leave the top radio button regarding recursively creating groups checked).

After adding the files to the bundle in this manner, you have to update your project's Information property list (.plist) file. If you highlight the .plist file in the Groups & Files view (you will find it in the Resources folder), several rows of data are displayed in the Editor view. Select the Icon file row to highlight it, and some small up and down arrows will appear. Click on them and you'll get a dropdown menu of all of the data types that can be specified in the .plist file for your app. Few of these are used or changed for a NimbleKit-based application, but in this case just change Icon file to Icon Files. Now you can add more than one icon to your application bundle.

After you do this, you'll get a large gray disclosure arrow to the left. Click on it and you'll see the first icon row, Item 0 (**Figure 3.7**). Enter Icon.png as the first icon file. For each additional icon file that you're bundling into your app, click the + button to the right of the row and you'll get a new row with one item number larger. Repeat until all your icon files are listed (but don't include the 512 × 512 iTunes icon, which is not bundled into the project file).

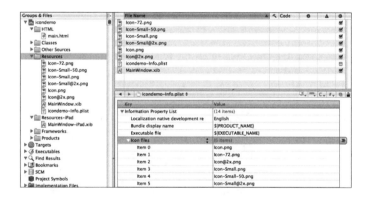

3.7 Your app icons in the project's Resources folder (Groups & Files and Detail views) and as listed in the .plist file (Editor view).

Designing a launch graphic

The launch graphic for an app is not necessarily what you think it is. It is not meant to be a title screen for your app (though if you design it this way, Apple might still accept it). But Apple recommends including a launch graphic for user experience reasons, not marketing reasons. And here's why.

It usually takes a few seconds for an app to launch and get its act together, so Apple has made a provision for a launch graphic to display during the app's launch process. But here's the catch: You can't control how long the graphic displays, so you can't specify that it show for a certain period of time (for example, if there's a line or two of text on it for the user to read). And for a smaller or more quickly loading app, the load time might be pretty fast, so the launch graphic flashes only briefly on the screen.

Still, the graphic is intended to serve a particular purpose and provide a smoother transition between the home screen and the app. This prevents the user from being jolted from the home screen to a completely blank screen and then to your app. You're welcome to read a more detailed explanation of this in Apple's human interface guidelines for the iPhone and iPad, but the main takeaway is this: Design the launch graphic to look as much like the initial view of the app as you can, minus any content, and this will help your app reinforce Apple's preferred iOS user experience.

To better understand this, let's look at what Apple has done with one of its own apps. Although we use some of these apps all the time, the very effectiveness of their design details is often lost on us. So we need to look a little more closely and stop the action. It's like dissecting a frog in science class. Sure, we know that a frog can hop far, but we can't see all the reasons why until we have a closer look at its legs! And, er, we need to stop the frog in order to take an even closer look inside at the bones and muscles used to hop.

But in this case, you don't need to kill an app to dissect it. **Figure 3.8** shows some screenshots of Apple's Clock app on the iPhone to illustrate my point.

3.8 The launch graphic (left) and alarm screen (right) in the Clock app.

Note how the launch graphic incorporates the pinstripe background of the app as well as the tab bar navigation at the bottom of the screen. And after Clock finishes launching on my iPhone and I go to my Alarm view, I see the image shown on the right in Figure 3.8. The two alarms that I've set appear, the Alarm tab is highlighted, and a few items (an Edit button, the Alarm title, and a + button for adding a new alarm) appear in the navigation bar.

The launch graphic eased me into this view very smoothly—it's like a three-page flip book, and when it's designed according to Apple's specs it's just as convincing as a well-drawn cartoon animation.

As you might have guessed, the launch graphic also needs to be sized to the device. **Table 3.2** shows the specs.

TABLE 3.2 Size (in pixels) and name specifications for iOS launch graphics

ORIENTATION	IPHONE / IPOD TOUCH	IPHONE 4	IPAD
Portrait	320 × 480 (Default.png)	640 × 940 (Default@2x.png)	768 × 1004 (Default-Portrait-ipad.png)
Landscape	n/a	n/a	1024 × 748 (Default-Landscape-ipad.png)

After adding your image files to your Resources folder, you need to modify your property list file just like you do when you add your app icons. Xcode allows you to specify launch images for both the iPhone and iPad platforms. Add the Default.png and Default-Portrait-ipad.png images to the list. Following the naming conventions shown in Table 3.2 allows your app to automatically load the iPhone 4 image when the app runs on an iPhone and allows the iPad landscape image to automatically load when in landscape mode on the iPad. When you're all done, **Figure 3.9** (on the next page) displays what you should see.

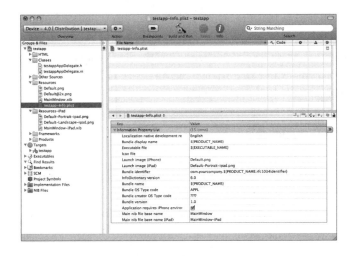

3.9 The Groups & Files pane shows the launch image files and the info. plist file shows the launch images specified for iPhone and iPad.

Setting the app version

Further down the rows in the Information Property List (-Info.plist) file is the bundle version (1.0, as shown in Figure 3.8). When you design your app, decide on a versioning nomenclature that will best suit the lifecycle of your app and its content. Will the app's content be updated regularly and, if so, how often? Or, if your app is for a client, does the client expect frequent design updates for any particular reason? As we know, a lot of retail product packaging is redesigned regularly, so if your app is tied to a product or service that encounters frequent brand updates, your app might have these same encounters and merit updates such as 1.2, 1.3, and 1.4 instead of 1.0, 2.0, 3.0, and 4.0.

The bundle version is an internal record of the app's version number that resides in the app bundle itself, whereas you also submit a version number via iTunes Connect with your packaged app binary (more on that in Chapter 12). I suggest that you always make these numbers match to avoid any confusion (either your own or Apple's), just in case having them not match could cause a problem.

Testing and building your app binary

Note that the app building process—part of both provisioning and testing your app on a device, as well as submitting to Apple for distribution via iTunes—requires you to be a registered and paid member of the Apple Developer Program. It also requires you to have purchased a license for NimbleKit. These processes are described in greater detail in Chapter 12.

But before you provision and test on a device, you can use the iOS SDK's Simulator for free. So we'll learn the settings via the Simulator process, and change them along the way for provisioning to a device and then eventually for submission to Apple.

Testing on Simulator

Testing on Simulator used to be a very straightforward process, and fortunately it's still one of the easier steps of iOS app design. But as with the other complications that the expansion of the iOS device family has caused, the three device types have made testing apps on Simulator a little less elegant than it used to be.

To launch and install the new NimbleKit app on Simulator, go to the Overview drop-down menu in the left corner of the Xcode project window. It is the extra-wide drop-down menu that should read **Simulator – 4.0 | Debug** followed by the name of the app you created at the beginning of this chapter. And if you've since closed that one or got rid of it, just create a new NimbleKit app project again.

When you click the Overview menu (**Figure 3.10**), make sure that Simulator is checked in the top group of options. In the next group, Active Configuration, Debug should also be selected. The next group, Active Target, will have only one item in it: the name of your app project file.

3.10 Xcode's Overview menu showing device and active configurations.

The next group, Active Executable, is where you specify which device you want Simulator to emulate when you begin testing your app. For our purposes, make sure that iPhone Simulator 4.0 is selected. Disregard Active Architecture; it is set automatically.

Here's where it gets a little exciting already: You'll install your first app on the Simulator, even though you haven't designed anything yet. Just for practice!

Doing this is very easy: Just click on the Build and Run icon in the middle of the Xcode project toolbar. Look for the clever Build and Run icon (a hammer and a green play button). The result looks like **Figure 3.11** (on the next page).

3.11 A new NimbleKit app after installation on the iPhone Simulator.

If you want to tinker with the Simulator a bit, do so now. It behaves just like a real iPhone with a few exceptions, like not having several of the standard apps (including Camera, Maps, Calendar, and so on). Also note that because there is no App Store app, you can't purchase and install other apps on Simulator—you can only install them through Xcode.

After playing with the screen version of the iPhone, try testing your universal app in the iPad Simulator. Close the Simulator, toggle back to Xcode, then use the Overview menu to specify iPad Simulator 3.2 as the Active Executable instead. Click the trusty Build and Run button again and you'll see a slightly different view of the new NimbleKit app, this time in the iPad Simulator (**Figure 3.12**).

3.12 A new NimbleKit app after it's been installed on the iPad Simulator.

The default view of iPad Simulator is 50 percent. You can change the view by clicking the Window menu and selecting Scale, where you can then choose 100 percent.

The last device you can simulate is iPhone 4, which I've saved for last because it is the easiest one to do. In fact, you've already installed your app on the iPhone 4 Simulator—congratulations! (You did this when you installed it on the iPhone Simulator.)

To view it there, go to the Simulator's Hardware menu, select Device, and then select iPhone 4. The Simulator's iPhone screen transforms into the iPhone 4 screen. Another reason I saved this for last is that, oddly, the iPhone 4 screen looks a lot more like the iPad screen you just saw than it does the iPhone screen. It lacks the chrome surround of the actual device. Don't fret about this, that's just what Apple did.

If the white app icon isn't displaying on the home screen, just flick over to the second home screen, where it was likely installed.

So congratulations again—now you can test apps on three devices in the Simulator! But testing there is never enough. Once you've designed an app of your own for distribution or sale, you need to test it on at least one actual iPhone, iPod touch, or iPad (and preferably more than one)—not just for extra quality assurance and control, but in some cases what you're testing can't fully function in Simulator. An example of this is an application with a map link that opens in the Maps application; there is no Maps app in Simulator, so testing that function on an iPhone is the only alternative.

Provisioning and testing on a device (debugging)

The first steps to creating the app binary for testing on a device are obtaining developer and provisioning certificates from the iOS Provisioning Portal, a section of developer.apple.com. This requires you to be a paid member of the iOS Developer Program (remember, it has an annual cost of $99), so I've put that section in the last chapter of this book because I wouldn't anticipate you taking this step until you've honed your app design skills. Whenever you decide to take that step, follow the directions in Chapter 12 and then return here to proceed with testing an app on a device.

After obtaining and installing your developer and provisioning certificate, you need to verify that Xcode has assigned the correct values to the app's build settings. The reason for this is in case the Development Provisioning Assistant, an online tool in the Provisioning Portal on developer.apple. com, hasn't properly set everything for you. Should that happen, the following directions will help you troubleshoot the situation and proceed with provisioning and device testing.

There are three areas in the project where build settings are defined and, mysteriously, sometimes they all need to be double-checked because setting them correctly in one place doesn't guarantee that they'll be correct in all three areas.

No, I'm not making this up. Furthermore, this has, at times, been a source of some of my most frustrating moments. I swear that I've set them all correctly and tried to test on my device, only to then return to see that a setting has changed. So I rub my eyes in disbelief, change the setting again, and *then* it works.

This is simply how it goes with Xcode at times. So be patient, take a deep breath, and recheck your settings whenever you run into a dead end.

The first place to check your settings is in the Overview menu in the upper left corner of the main Xcode project window. As covered previously in this chapter, make sure that this is set to

- Device
- Active Configuration: Debug (same as installing in Simulator)
- Active Target: the name of your app (there should only be one option here)
- Active Executable: the name of your device (again, there should only be one option)

The build settings can be reached via the information window of two different files:

- The Project file
- The Target file

You reveal these settings by selecting them one at a time and clicking the blue info button in Xcode's settings (or by just right-clicking the file and selecting Get Info) (**Figure 3.13**).

3.13 The Project and Target files. (On screen, these will be shown selected in blue.)

When you do this, you'll get a large Project Info window with four tabs: General, Build, Configurations, and Comments. There are a ton of settings in these tabs but, fortunately, there are very few that you need to understand or change. The settings for our current purposes are found under the Build tab (**Figure 3.14**):

- Configuration: set to Debug

- Base SDK: set to the most current SDK (iPhone Simulator 4 in this example)

- Code Signing Identity: set to any iPhone device on the left and Automatic Profile Selector | iPhone Developer on the right

- Based On: set to NKit

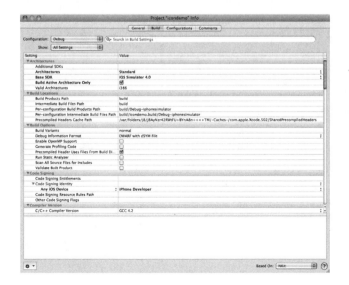

3.14 The Build tab of the Project Info window.

Now the trick of this is—and I kid you not—you can't just submit these values for the project file and expect them to always show up in the app file as well. (Though this can sometimes happen—and even often.) I'm afraid it's a hard lesson to learn about how Xcode works, and sometimes these details even vary a bit from app to app. However, I do believe that the Development Provisioning Assistant in the Provisioning Portal has made this more consistent.

So at any point when you click Build and Run and your app is not installed on your device, double-check these same settings in the Target Info window too. Make sure everything is set as outlined above.

Building and submitting (distributing)

When you submit an app to Apple for review and placement in the iTunes App Store, the process isn't that different from using Xcode to install the app on a device. And again, the details of building and submitting make more sense at the end of this journey, so they are covered in Chapter 12.

PURCHASING AND ACTIVATING YOUR NIMBLEKIT LICENSE

In addition to needing a valid Distribution Provisioning Profile for your app, you'll need a licensed version of NimbleKit. At the time of this writing, a NimbleKit license requires a one-time fee of $99. After purchasing the license, you'll receive a 16-digit serial number.

After receiving the number, go to the Classes folder in your app's project. Yes, as previously mentioned, this is one of the areas that is normally off-limits to you. But this is one time you're allowed to go here. Open the folder, then double-click the AppDelegate.m file. You'll see a bunch of Objective-C code that you would normally not touch.

Look for this line of code:

```
Nimble *nimble = [[Nimble alloc] initWithRootPage:@"main.html"
window:window serial:@""];
```

Copy and paste your NimbleKit serial number between the double quotes, so you see this (only the Xs below represent your actual serial number):

```
Nimble *nimble = [[Nimble alloc] initWithRootPage:@"main.html"
window:window serial:@"XXXX-XXXX-XXXX-XXXX"];
```

Then save, exit, and do not tamper with an Objective-C file again. You are now licensed to use NimbleKit and distribute NK-based apps to iTunes!

After creating a Distribution Provisioning Profile on the iOS Provisioning Portal and downloading it to your computer, drag it onto your Xcode app icon to install.

Open your app project and select the Project file at the top of the Groups & Files pane. View its information and then click the Configurations tab. Duplicate the Release configuration and then rename the copy Distribution. You have now completed one of the strangest steps of distributing an app. Weird, huh?

Fortunately from here on out, it feels a lot like provisioning the app for testing.

Set the Overview drop-down menu's settings as follows:

- Destination: set to Device
- Active Configuration: set to Distribution

Set the Project and Target files settings in their info windows, under the Build tab:

- Configuration: set to Distribution (**Figure 3.15**)
- Base SDK: set to the most current SDK (iPhone Device 4 in this example)
- Code Signing Identity: set to any iPhone device on the left and Automatic Profile Selector | iPhone Distribution on the right
- iPhone OS Deployment Target: iPhone OS 3.0
- Based On: set to NKit

3.15 The Overview menu showing Distribution settings.

Have you verified that these are your settings for both the Project file and Target file? Good. And remember that if you encounter a build error, double-check all of these again—sometimes Xcode can be very finicky and seemingly even a bit scatterbrained.

Now that your app settings are in order, it's time to build your app binary!

Before doing so, go to the Build menu and select Clean All Targets. This should ensure that anything generated during Simulator and device testing won't interfere with the generation of a clean app binary.

Then—drumroll, please—return to the Build menu and click Build. Nope, this isn't a typo: There's a Build command in the Build menu!

Xcode should then take a moment to process the binary, and if you get no warnings or errors, this process should result in the file you will then submit to iTunes Connect for approval by Apple. The file is an .app file that is located in the Distribution-iPhoneos subdirectory. When you find it, select it and compress it (right-click or Ctrl-click, and then choose Compress). You should see what is shown in **Figure 3.16**.

3.16 Your app binary, prepared for distribution. Don't forget to compress it!

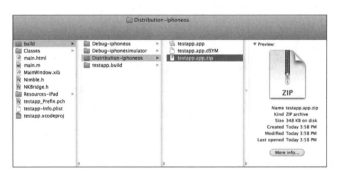

At this point you've learned the basics of Xcode and Simulator, the two main tools in the iOS SDK. To complete the examples in this book and design your own apps, you'll continue to add files to the HTML directory, edit them in the Editor pane, and test your work in Simulator.

Summary

In this overview of Xcode and Simulator, you have learned how to

- Start new NimbleKit app projects in Xcode for iPhone and iPod touch, iPad, or for all iOS devices (universal app).
- Use the Groups & Files, Detail, and Editor panes to manage files, app settings, and edit code.
- Avoid the NimbleKit framework files so you don't inadvertently disable the Objective-C code (which allows your project to run natively and, eventually, be approved by Apple).
- Name (and later rename) your app.
- Design and add icon and launch graphics to your app that meet Apple's specifications and user experience guidelines.
- Set the app version.
- Build your app binary for Simulator and device testing.
- Build and compress your app binary so you can submit it to iTunes Connect for Apple approval.

Now that you know the ins and outs of using Xcode and creating a project, it's time to start learning about iOS apps themselves! You'll begin by learning about some standard iOS interface elements.

4 THE IOS INTERFACE AND USER EXPERIENCE

The iPod touch and iPhone screens—
they're so small! How is it possible to
design a quality user experience on some-
thing as tiny as 320 × 480 pixels?

Coming from a large-screen environment where we are typically designing for 960 pixel, 1024 pixel, or even wider spaces, these mobile screens at first seem impossibly small. But millions of people are buying and using these devices, and they aren't doing it to torture themselves— we're seeing some great user experiences on these pocket computers. They can be fun, easy, and even delightful to use.

So what makes this possible? How can we learn from Apple's own apps and from other examples? This is what we'll be focusing on in this chapter, as well as some of Apple's suggestions for how to succeed on the small screen.

As we begin figuring out how to use this small piece of real estate, it's important to take a look at everything Apple has given us. This includes walking the walk and talking the talk, so you'll also be learning a new vocabulary.

Let's start with the real estate itself. **Figure 4.1** shows our 320 × 480 pixel screen (and note that in general discussions I will refer to the current iPod touch and iPhone 3GS screen size, though I will supplement with iPhone 4 and iPad sizes where appropriate):

4.1 The standard size iOS screen for iPod touch and iPhone (pre-4), where most iOS users are.

480 px

320 px

Screen is the most common term for the entirety of the view—it's not a window like on a desktop or laptop computer. Windows have close buttons, minimize, and enlarge buttons, and they reside on a desktop. But an iOS app screen just *is*—remember that this is a more immersive experience (consider screens in movie theaters), so thinking in terms of screens is a slight philosophical and practical change from designing less immersive, windows-based websites

Figure 4.2 shows the *New York Times* app and examples of the iOS interface elements that you'll learn how to implement in this chapter. The first element of the iOS screen you'll be introduced to is the status bar.

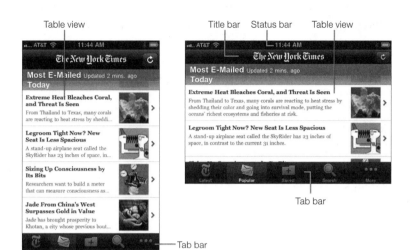

Table view   Title bar   Status bar   Table view

Tab bar

4.2 The *New York Times* iOS application in portrait and landscape orientations.

Tab bar

What is the status bar?

The iOS status bar (**Figure 4.3**), located at the top of the screen, grounds the device screen with some key information that we're all used to seeing. If it's an iPhone status bar, we see our cellular signal strength (or, as is often the case, our lack of signal strength!). And regardless of the device, we also see whether we're on a wireless internet connection and what its strength is.

4.3 The iOS status bar.

The local time is centered in the status bar. And a nice feature of this area is that tapping the center of the status bar is equivalent to a "Return to top of page" link on a website. It quickly scrolls the screen back to the top. This is a user interface feature that works in any app, so that's why all apps except video games tend to include the status bar in their design.

Finally, the rightmost element in the status bar is, of course, the battery charge indicator.

The default color of the status bar is silver, but it can also be set to black and black translucent. When using NimbleKit to trigger these native

Objective-C settings, here's how to change the status bar to black using JavaScript:

```
var application = new NKApplication();
application.setStatusBarStyle("black");
```

And for black translucent:

```
var application = new NKApplication();
application.setStatusBarStyle("blacktranslucent");
```

The default setting is, not surprisingly, "default." Furthermore, if you're not changing the color to black, don't even worry about including this snippet in your app—it's only necessary if you don't want the default silver appearance.

NOTE Using these examples in a NimbleKit-based app

If you want to try any of these code snippets as we go along, just open a new app in Xcode and choose the NimbleKit option as explained in the last chapter. Enter the code after **<script type="text/javascript" src="NKit.js"></script>** in the main.html file that is provided in the HTML subdirectory. I'm omitting opening and closing script tags in these samples to avoid repetition, so be sure to wrap these examples with **<script type="text/javascript">** and **</script>** like so:

```
<html>
<head>
<meta name = "viewport" content = "initial-scale = 1.0,
user-scalable = no" />

<script type="text/javascript" src="NKit.js"></script>
<script type="text/javascript">
<!-your NimbleKit JavaScript calls here -->
</script>

</head>

<body>
<!-your content here -->
</body>

</html>
```

And if you don't want to type them, download the code samples at iosapps.tumblr.com.

When the device is rotated and the app supports landscape orientation, the status bar expands to fill the new width. So when planning app screen designs, it's important to take aspects like this into consideration. To help you with your planning, I will provide portrait and landscape dimensions of iOS elements for all three devices. Let's start with the status bar. In **Table 4.1**, I use the design term *portrait* to mean the standard, upright device orientation and the term *landscape* to mean the rotated or "sideways" orientation.

TABLE 4.1 Dimensions of the iOS status bar (in pixels)

ORIENTATION	IPHONE/ IPOD TOUCH	IPHONE 4	IPAD
Portrait	320 × 20	640 × 40	768 × 20
Landscape	480 × 20	960 × 40	1024 × 20

Implementing the title bar

The next major element of the iOS user interface is the title bar (**Figure 4.4**), located immediately below the status bar in an app screen. The title bar is an absolutely critical element. As the name implies, it often confirms the title of an app upon launch. And the title bar functions much like the title of a web page in a site: It's the landmark that helps you maintain your bearings as you move from screen to screen through the app.

Screen Name

4.4 The iOS title bar.

The title bar has a default Apple color of blue-gray. Keeping the default color is a great design decision when you want to maximize the "nativeness" of your app.

On the other hand, setting a custom background color for the title bar is a nice design opportunity if you, your employer, or your client wants to brand an app a bit more. So I encourage you to consider this option, too. Setting a custom color can be a nice way to enhance the brand identity of an app while still keeping the familiar dimensions and gradient.

Let's walk through how to do this. The basic JavaScript is below (and the variables you can change are highlighted)

```
var navController = new NKNavigationController();
navController.setTitle("Title Here");
navController.setTintColor(127, 62, 152);
```

Note that the NimbleKit library item is called `NKNavigationController` (and not `NKTitleBar`). This is because as a user navigates into an app (via a table view or tab bar navigation), the `NKNavigationController` automatically adds native back buttons so users can return to where they started.

Regarding the title displayed in the bar, it has a limit, of course, and this depends on the device you're using—a screen title on an iPod touch or iPhone needs to be shorter than one on an iPad. But keeping them short is a good practice anyway. If you end up choosing a title that's too long, it will automatically be truncated with an ellipsis at the end.

The color setting uses RGB settings just like the CSS `rgb` color property does. So in the previous code example, I selected a violet hue to match a client's brand. I used Adobe Photoshop to open a file with the color palette in it, and used the Color Picker (**Figure 4.5**) to tell me what the RGB values were.

4.5 Using the Color Picker in Photoshop to find out a hue's RGB values.

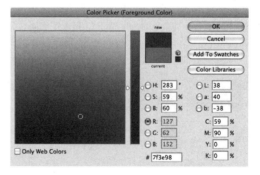

Note that the gradient effect comes standard with the NimbleKit title bar.

Table 4.2 shows the size of the title bar on the various iOS devices and in both portrait and landscape orientations for your planning purposes.

TABLE 4.2 Dimensions of the iOS title bar (in pixels)

ORIENTATION	IPHONE/ IPOD TOUCH	IPHONE 4	IPAD
Portrait	320 × 44	640 × 88	768 × 44
Landscape	480 × 44	960 × 88	1024 × 44

Designing with tab bars

The tab bar (**Figure 4.6**) gets us into more interesting iOS user experience territory, as it's one of the principal ways to navigate between different screens in a single application.

4.6 The iOS tab bar.

Tab bar with standard categories

Now you'll learn how to code the sample displayed in Figure 4.6. In this example, there are three tabs that navigate to three other pages. Each tab has two components: the titles (Favorites, Featured, and Top Rated) and an icon for each.

This first example does a lot of heavy lifting for you, because NimbleKit is able to call some of these native items that are built right into the operating system. In other words, both the titles and the icons come for free.

The basic JavaScript for this is (variables you can change are highlighted)

```
var tabController = new NKTabBarController();
tabController.setTabBarForPage("main.html", "", "1");
tabController.setTabBarForPage("two.html", "", "2");
tabController.setTabBarForPage("three.html", "", "3");
```

The way this works is that the page names for each tab are HTML file names. The second setting is for tab labels (covered in the next example), and the third setting refers to built-in categories and icons. The tab label settings in the previous code listing are left blank because they're not used when calling built-in tab categories (which are automatically labeled).

There are currently eleven built-in categories and icons (**Figure 4.7**).

4.7 The built-in tab
navigation categories.

1. Favorites   2. Featured   3. Top Rated   4. Recents

5. Contacts   6. History   7. Bookmarks   8. Search

9. Downloads   10. Most Recent   11. Most Viewed

NOTE The titles and icons are free, but not the functionality

What you get for "free" here are only the tab button titles and icons, not the associated functionality that would go with them. In other words, making a tab called Contacts doesn't automatically create a contacts screen for you.

Tab bar with custom categories

On the other hand, you may not need these categories. In that case, you'll want to assign your own labels to the tabs. In fact, you can even design your own icons and have them integrated into a tab bar. **Figure 4.8** shows a sample tab bar I've designed to demonstrate this.

4.8 Design your own
custom tab navigation.

Here's how the JavaScript differs for a custom version:

```
var tabController = new NKTabBarController();
tabController.setTabBarForPage("main.html", "Location",
"icon_globe.png");
```

```
tabController.setTabBarForPage("two.html", "Menu", "icon_
fork.png");
tabController.setTabBarForPage("three.html", "Tweets",
"icon_bird.png");
```

Take note of the items in the code that you must set:

- The first item is the page to which the tab navigates.
- The second item is the text label for the tab.
- The third item is the icon.

I've found that designing a PNG image that is 30 × 30 pixels with a transparent background works well for a tab icon. Of course, beyond those basic specifications, the scale and detail of the image will determine whether it is recognizable at that size. But as long as the PNG is a solid image overlaying a transparent background, the rest of the native effects (gradient, glow, and white and blue colors) are applied for free by NimbleKit and the operating system.

So what is the maximum number of tabs you can display in a tab bar navigation? For the iPod touch and iPhone it is five, and for the iPad it is eight. Except you can actually go beyond that—when you exceed the maximum number of tabs that the navigation will display, it automatically adds a More tab in the last position and puts everything else into a table view navigation (which you'll learn more about next) (**Figure 4.9**).

More button

Table View

4.9 An iPod touch/iPhone tab bar navigation with six categories, demonstrating how the More tab is automatically added by NimbleKit. Nice!

Table 4.3 shows the size specifications for the tab bar navigation, for purposes of laying out your screen design.

TABLE 4.3 **Dimensions of the iOS tab bar (in pixels)**

ORIENTATION	IPHONE/ IPOD TOUCH	IPHONE 4	IPAD
Portrait	320 × 49	640 × 98	768 × 49
Landscape	480 × 49	960 × 98	1024 × 49

Navigating with table views

Table views (**Figure 4.10**) are at the heart of iOS applications. They are ubiquitous, from Apple apps that come installed on iOS devices (Settings, Mail, Contacts) to any app that delivers content to people.

This section explores some of the types of table views that are possible, and dives into how to implement them in your apps.

4.10 A plain table view navigation.

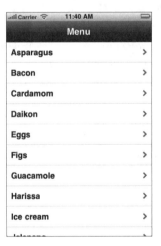

Plain table view

The plain table view navigation shown in Figure 4.10 is as native as it gets: it's a plain table with no adornments, and navigates to additional screens in the app. In fact, it's exactly like the table view shown in Apple's iPod application (**Figure 4.11**).

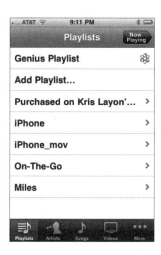

4.11 A plain table used to show playlists in the iPod application on the iPhone.

To code the table, use NimbleKit's NKTableView function. Here's the JavaScript that calls the **NKTableView** library item shown in Figure 4.10:

```
var tableView = new NKTableView();
tableView.init(0, 0, 320, 440, 'plain');
tableView.insertRecord("Asparagus", "", "", "0", "",
"navController.gotoPage('1.html')");
tableView.insertRecord("Bacon", "", "", "0", "",
"navController.gotoPage('2.html')");
tableView.insertRecord("Cardamom", "", "", "0", "",
"navController.gotoPage('3.html')");
tableView.insertRecord("Daikon", "", "", "0", "",
"navController.gotoPage('4.html')");
tableView.insertRecord("Eggs", "", "", "0", "",
"navController.gotoPage('5.html')");
tableView.insertRecord("Figs", "", "", "0", "",
"navController.gotoPage('6.html')");
tableView.insertRecord("Guacamole", "", "", "0", "",
"navController.gotoPage('7.html')");
tableView.insertRecord("Harissa", "", "", "0", "",
"navController.gotoPage('8.html')");
tableView.insertRecord("Ice cream", "", "", "0", "",
"navController.gotoPage('9.html')");
tableView.insertRecord("Jalapeno", "", "", "0", "",
"navController.gotoPage('10.html')");
tableView.show();
```

When you're making this, you're creating the instance of `NKTableView` called `tableView`, defining how much of the screen it takes up—in this case, the entire screen except the status and title bars—and specifying that it's a plain table view (the alternative is grouped, which you will explore next).

After this, use `insertRecord` to add the desired number of rows (it will scroll, so there really isn't a limit that I'm aware of). The six parameters available are:

- title
- subtitle
- left image
- section number
- right image
- callback

NOTE Table view usability and information hierarchy:
How many rows are too many?

I'm not aware of a limit to the number of rows in a table view, but that doesn't necessarily mean there is no practical limit. If you're considering having a ton of rows in a table view, you might be better off using a grouped table view or even multiple table views (that is, a table view that links to screens with additional table views). Consider the usability and information hierarchy concerns of how you design with this interface element.

In a basic table view, you leave the subtitle and image parameters empty (images are covered in the next example). The section parameter is used to specify sections in a grouped table view; here it is set to `0` (quirk alert—otherwise the row will not display!). Finally, the callback here is scripted to put the `navController` function into motion and navigate to screens that would tell people more about these various foods.

Table view with images

Table views can also be designed with images in them. You've seen this in screens like the album view in the iPod app (**Figure 4.12**).

4.12 The album view in Apple's iPod app on the iPhone.

To achieve this result, use NimbleKit's **NKImage** control and assign the instance a name, in this case, **image**. Here's a line of JavaScript for this:

```
var image = new NKImage();
```

Then you need some images to pull into your table view rows. In a normal-sized row, the height is 43 pixels; you can either size your images to this or let the table view resize them for you. Prepare 72 dpi PNG images and crop them to squares exactly the way you want them. For this sample, I've found images of asparagus and bacon (**Figure 4.13**), and have deliberately left them different sizes—asparagus.png is 100 pixels square and bacon.png is 183 pixels.

4.13 Asparagus! Bacon!

Then you need to add the files to your Xcode project (go to Project and choose Add to Project) so they're in the HTML group (**Figure 4.14**).

4.14 The asparagus.png and bacon.png files, now in the Xcode project.

If you begin with the same code you used in the plain table view section, the changes are highlighted as follows for activating the **NKImage** control within the table view rows:

NOTE

Where sample content
includes several items or
longer passages, code
may be omitted in the
sample to keep it short.
The ellipsis (...) shows
where this happens, as
shown on this page when
items 3–10 of the table
view are not repeated in
the code.

```
var tableView = new NKTableView();
tableView.init(0, 0, 320, 440, 'plain');
image.loadFromBundle("asparagus.png");
tableView.insertRecord("Asparagus", "", image, "0", "",
"navController.gotoPage('1.html')");
image.loadFromBundle("bacon.png");
tableView.insertRecord("Bacon", "", image, "0", "",
"navController.gotoPage('2.html')");
...
tableView.show();
```

So the main.html file, after adding the image modifications, is

```
<!DOCTYPE html>
<html>
<head>
<meta name = "viewport" content = "initial-scale = 1.0,
user-scalable = no" />

<script type="text/javascript" src="NKit.js"></script>

<script type="text/javascript">
var navController = new NKNavigationController();
navController.setTitle("Menu");
navController.setTintColor(0, 63, 78);

var image = new NKImage;

var tableView = new NKTableView();
tableView.init(0, 0, 320, 440, 'plain');
image.loadFromBundle("asparagus.png");
tableView.insertRecord("Asparagus", "", image, "0", "",
"navController.gotoPage('1.html')");
image.loadFromBundle("bacon.png");
tableView.insertRecord("Bacon", "", image, "0", "",
"navController.gotoPage('2.html')");
...

tableView.show();

</script>
</head>
<body>
</body>
</html>
```

This changes the appearance of the top two rows, as shown in **Figure 4.15**.

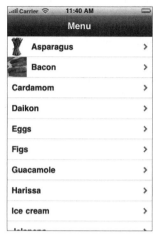

4.15 The asparagus and bacon images displayed in their table view rows.

I'm certainly not advocating that you only add photos to some of the rows in a table view. I just didn't want to hunt down images of all those foods for this example!

Grouped table view

The other main category of table view is a grouped view (**Figure 4.16**). These are also quite common, such as this one from the Settings app on an iPhone.

4.16 A grouped table view from the iPhone's Settings app.

The grouped table view works well if you have items that relate to each other. With grouped views, you can also relate them visually. Let's build one of our own to see how it works.

Figure 4.17 shows what you'll be making in this example.

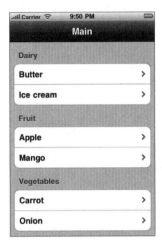

4.17 A grouped table view showing foods in food groups.

And the following code shows the JavaScript that makes it. I've highlighted the code differences between the grouped and plain examples:

```
var tableView = new NKTableView();
tableView.init(0, 0, 320, 440, 'grouped');

tableView.insertCategoryNamed('Dairy');
tableView.insertRecord("Butter", "", "", "0", "",
"navController.gotoPage('1.html')");
tableView.insertRecord("Ice cream", "", "", "0", "",
"navController.gotoPage('2.html')");

tableView.insertCategoryNamed('Fruit');
tableView.insertRecord("Apple", "", "", "1", "",
"navController.gotoPage('3.html')");
tableView.insertRecord("Mango", "", "", "1", "",
"navController.gotoPage('4.html')");

tableView.insertCategoryNamed('Vegetables');
tableView.insertRecord("Carrot", "", "", "2", "",
"navController.gotoPage('5.html')");
```

```
tableView.insertRecord("Onion", "", "", "2", "",
"navController.gotoPage('6.html')");
```

```
tableView.show();
```

As you can see, you need to set the type in `tableView.init` to `grouped` and then add the `.insertCategoryNamed` lines for each group. The thing that's a bit strange, however, is that those lines themselves do not group the items. What actually groups them is the fourth parameter of `.insertRecord`. So, in this example, the three groups are numbered **0**, **1**, and **2**.

Unlike the previous elements, the table view does not have a set dimension in both width and height because the height depends on the number of rows you use. As shown in **Table 4.4**, table views only have set widths.

TABLE 4.4 **Widths of the iOS table view (in pixels)**

ORIENTATION	IPHONE/ IPOD TOUCH	IPHONE 4	IPAD
Portrait	320	640	768
Landscape	480	960	1024

Summary

In this chapter, you learned how to design some fundamental elements that will make your apps look and behave in a way that's familiar to owners of iOS devices. Now you can

- Plan your screen layouts better by understanding what core interface elements are called and what size they are.
- Use the status bar as a "return to top of page" button.
- Implement and customize the color of the title bar.
- Use the tab bar navigation when you have a smaller screen count for your app.
- Use the table view navigation when you have a longer list of content categories, a larger number of secondary screens, or the need to group navigation items in your app.
- Add images to a table view navigation.

Next, you'll learn how to incorporate and style text and image content.

5 FOCUS ON APP CONTENT: TEXT AND IMAGES

Words and images: our primary forms of visual communication.

And our most important content for iOS applications.

The basis for any content-based mobile application is, quite naturally, content. The forms in which we communicate primarily involve words, pictures, and often both. Fortunately, all of the understanding and experience that web designers have accumulated about structuring and formatting text and image content for the web translates very well to iOS apps designed with NimbleKit.

But before you think that working with text and images is as simple as wrapping paragraphs in `<p>` tags and properly sizing and formatting photos, let's consider how HTML allows us to design content semantically—that is, with a structural hierarchy and logic. We'll do so by considering some principles that should be familiar already, and focus on them through the unique lens of iOS devices and iOS app examples to sharpen the already-strong instincts we've honed by designing standards-based websites.

Structuring text

To begin exploring best practices for structuring text for iOS apps, we need a suitable sample of text that has some kind of structure. A good example of this in the public domain is the United States Bill of Rights.

There are a few different ways to go about structuring this content with HTML—let's explore them to see how they affect the way the text is displayed.

Ordered list

One way to structure the Bill of Rights, in light of there being ten of them that need to appear in a particular order, is to make an ordered list. The semantic meaning would be conveyed clearly, and one might argue that this method is also the most convenient—in this case, NimbleKit's web view does the numbering for us just like any other browser engine. Such a list would begin as follows:

```
<h1>The United States Bill of Rights</h1>
<ol>
<li>Congress shall make no law respecting an establishment
of religion, or prohibiting the free exercise thereof; or
abridging the freedom of speech, or of the press, or the
right of the people peaceably to assemble, and to petition
the Government for a redress of grievances.</li>

<li>A well regulated Militia, being necessary to the
security of a free State, the right of the people to keep
and bear Arms, shall not be infringed.</li>…

{amendments 3-10}

</ol>
```

The results are rendered as shown in **Figure 5.1** (on page 70).

UNITED STATES BILL OF RIGHTS

1st amendment: Congress shall make no law respecting an establishment of religion, or prohibiting the free exercise thereof; or abridging the freedom of speech, or of the press, or the right of the people peaceably to assemble, and to petition the Government for a redress of grievances.

2nd amendment: A well regulated Militia, being necessary to the security of a free State, the right of the people to keep and bear Arms, shall not be infringed.

3rd amendment: No Soldier shall, in time of peace be quartered in any house, without the consent of the Owner, nor in time of war, but in a manner to be prescribed by law.

4th amendment: The right of the people to be secure in their persons, houses, papers, and effects, against unreasonable searches and seizures, shall not be violated, and no Warrants shall issue, but upon probable cause, supported by Oath or affirmation, and particularly describing the place to be searched, and the persons or things to be seized.

5th amendment: No person shall be held to answer for a capital, or otherwise infamous crime, unless on a presentment or indictment of a Grand Jury, except in cases arising in the land or naval forces, or in the Militia, when in actual service in time of War or public danger; nor shall any person be subject for the same offence to be twice put in jeopardy of life or limb; nor shall be compelled in any criminal case to be a witness against himself, nor be deprived of life, liberty, or property, without due process of law; nor shall private property be taken for public use, without just compensation.

6th amendment: In all criminal prosecutions, the accused shall enjoy the right to a speedy and public trial, by an impartial jury of the State and district wherein the crime shall have been committed, which district shall have been previously ascertained by law, and to be informed of the nature and cause of the accusation; to be confronted with the witnesses against him; to have compulsory process for obtaining witnesses in his favor, and to have the Assistance of Counsel for his defence.

7th amendment: In suits at common law, where the value in controversy shall exceed twenty dollars, the right of trial by jury shall be preserved, and no fact tried by a jury shall be otherwise re-examined in any Court of the United States, than according to the rules of the common law.

8th amendment: Excessive bail shall not be required, nor excessive fines imposed, nor cruel and unusual punishments inflicted.

9th amendment: The enumeration in the Constitution, of certain rights, shall not be construed to deny or disparage others retained by the people.

10th amendment: The powers not delegated to the United States by the Constitution, nor prohibited by it to the States, are reserved to the States respectively, or to the people.

Source: http://en.wikipedia.org/wiki/United_States_Bill_of_Rights

5.1 The Bill of Rights as an ordered list.

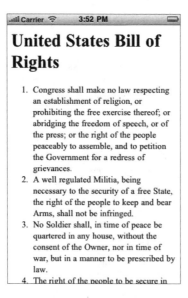

Even unstyled (that is, without any CSS giving it additional style), an unordered list with a heading above it conveys a lot of structure and meaning. And it comes with the "free styling" of the auto-numbering, so that's kind of cool. And by styling the `<li>` tags, you could give this structure quite a bit more artistic flair.

One downside to this content structure is having the entire list in one screen. For an app that is only about the Bill of Rights, this might be overly simplistic; for an app that is about the entire Constitution and its Amendments, it might be entirely appropriate.

Definition list

Another way to structure this content is to place it in a definition list. I'm a fan of definition lists because I like the added built-in semantic meanings—I just think it's particularly clever that the tags define the list with a `<dl>`, the term with a `<dt>`, and the definition or detail with a `<dd>`. It has a poetic clarity. So in this case, the list of amendments begins as follows:

```
<h1>United States Bill of Rights</h1>
<dl>
<dt>1<sup>st</sup> Amendment</dt>
```

```
<dd>Congress shall make no law respecting an establishment
of religion, or prohibiting the free exercise thereof; or
abridging the freedom of speech, or of the press, or the
right of the people peaceably to assemble, and to petition
the Government for a redress of grievances.</dd>

<dt>2<sup>nd</sup> Amendment</dt>
<dd>A well regulated Militia, being necessary to the
security of a free State, the right of the people to keep
and bear Arms, shall not be infringed.</dd>…

{amendments 3-10}

</dl>
```

The results are rendered as shown in **Figure 5.2**.

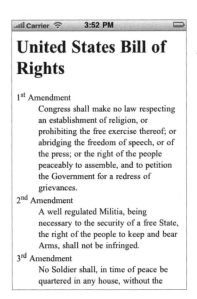

5.2 The Bill of Rights as a definition list.

As shown in Figure 5.2, even the unstyled definition list has a bit of style to it, which I also really like. The browser engine automatically gives the definition a jaunty indent, so the list is eminently readable. And naturally, you could add to this with CSS by giving it a background, or even undoing the indent—there are countless ways to style this list. But on its own, it's a nice serviceable starting point. And more importantly: If this app had a companion website, you could style both versions of your amendment content with the same semantic markup. I like knowing that when I start with nice web markup, my app markup can follow suit and, paired with NimbleKit, yields a nice product.

As with the ordered list version, though, the definition list still might not be your preferred way to deliver this content unless you're designing an app about something more than the Bill of Rights. A one-screen app is not especially compelling and doesn't give you much opportunity to leverage native iPhone user experiences.

Table view

A third way to structure this content is to design a table view that goes to each amendment separately. For an app that focuses exclusively on the Bill of Rights, this is arguably the most native way to structure the content. It delivers a more genuine and interactive iOS user experience.

To code the table, use NimbleKit's `NKTableView` function that we learned about in the previous chapter. Here's the JavaScript that calls the `NKTableView`:

```
var billView = new NKTableView();
billView.init(0, 0, 320, 440, 'plain');
billView.insertRecord("1st Amendment", "", "", "0", "",
"navController
   .gotoPage('1.html')");
billView.insertRecord("2nd Amendment ", "", "", "0", "",
"navController
   .gotoPage('2.html')");
billView.insertRecord("3rd Amendment ", "", "", "0", "",
"navController
   .gotoPage('3.html')");
billView.insertRecord("4th Amendment ", "", "", "0", "",
"navController
   .gotoPage('4.html')");
billView.insertRecord("5th Amendment ", "", "", "0", "",
"navController
   .gotoPage('5.html')");
billView.insertRecord("6th Amendment ", "", "", "0", "",
"navController
   .gotoPage('6.html')");
billView.insertRecord("7th Amendment ", "", "", "0", "",
"navController
```

```
   .gotoPage('7.html')");
billView.insertRecord("8th Amendment ", "", "", "0", "",
"navController
   .gotoPage('8.html')");
billView.insertRecord("9th Amendment ", "", "", "0", "",
"navController
   .gotoPage('9.html')");
billView.insertRecord("10th Amendment ", "", "", "0", "",
"navController
   .gotoPage('10.html')");
billView.show();
```

The results are rendered as shown in **Figure 5.3**.

5.3 The Bill of Rights as a table view.

Next is the HTML for the first amendment screen (1.html)—it's quite brief:

```
<html>
<head>
<meta name = "viewport" content = "initial-scale = 1.0,
user-scalable = no" />

<script type="text/javascript" src="NKit.js"></script>
<script type="text/javascript">
var navController = new NKNavigationController();
```

```
navController.setTitle("1st Amendment");
</script>

</head>

<body>

<p>Congress shall make no law respecting an establishment
of religion, or prohibiting the free exercise thereof; or
abridging the freedom of speech, or of the press, or the
right of the people peaceably to assemble, and to petition
the Government for a redress of grievances.</p>

</body>
</html>
```

This yields the result shown in **Figure 5.4**.

5.4 The first amendment as a separate app screen, called from the table view.

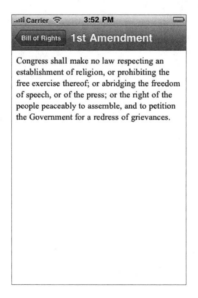

This is not yet styled and is kind of boring, so it could use a fair amount of CSS love. But you get the idea.

So that's another way to structure this Bill of Rights content. It results in a nicely styled table view, with great NimbleKit-powered transitions.

Integrating social content

Compared to offering free content via a web app or mobile-enabled website, designing and packaging a native iOS app for distribution in the iTunes App Store presents one notable hurdle: How can you or a client easily update content for app users without updating the entire application? Fortunately, there are a few tricks for working around this obstacle, so you don't necessarily have to submit an updated app (and then wait for approval) whenever you have new content.

The idea in this example is to pull Twitter status updates into an app screen, just as many people do with their blogs or other websites where they want their tweets displayed in a column adjacent to their other site content. We're basically using Twitter as a content management system. To accomplish this, people rely on JavaScript; one of the most common solutions is to use a script provided by Twitter itself, located at http://twitter.com/javascripts/blogger.js.

USING NIMBLEKIT TO CHECK FOR INTERNET CONNECTIVITY

When accessing content from the Internet, Apple wants apps to be able to check for Internet connectivity and warn users when there is no wireless or cellular connection. To do this, an alert sheet (**Figure 5.5**) is used to convey the message (`NKAlert`):

5.5 An alert sheet that tells the user there is no Internet connectivity.

and two other library items are used to check for the connectivity:

- `NKIsInternetAvailableViaWifi`
- `NKIsInternetAvailableViaCellularNetwork`

(continues on next page)

Below is a script that I recommend using in the head of all NimbleKit app HTML pages that need to pull content from the Web:

```
// check for internet connectivity
function checkForInternet()
{
  var isInternetAvailable = false;
  if (NKIsInternetAvailableViaWifi()==1)
  {
    isInternetAvailable = true;
  }
  else
  {
    if (NKIsInternetAvailableViaCellularNetwork()==1)
    {
    NKAlert("Info", "Internet is available only via cellular network,
    carrier fees may apply.");
    isInternetAvailable = true;
    }
  else
    {
    NKAlert("Error", "An internet connection is required to access
    content
    for this function.");
    isInternetAvailable = false;
    }
  }
  return isInternetAvailable;
}
if (checkForInternet())
{
  // load content from the web
}
```

This script will help us out with our task, too. We just need to make sure that our app opens links in Mobile Safari. If we don't do this, the links will open in the app itself. That would be rather tacky and present us with some navigational challenges because the linked page would become the new state of this view. Yet you don't have a back button like you would in a browser. So you need to call on `NKOpenURLInSafari`, which tells your links to open in Mobile Safari rather than in the app itself.

The modified blogger.js JavaScript file looks like this, with the `NKOpenURLInSafari` addition highlighted to show you where it goes:

```
function twitterCallback2(twitters) {
  var statusHTML = [ ];
  for (var i=0; i<twitters.length; i++){
    var username = twitters[i].user.screen_name;
    var status = twitters[i].text.replace(/
    ((https?|s?ftp|ssh)\:
\/\/[^"\s\<\>]*[^.,;'"">\:\s\<\>\)\]\!])/g, function(url)
{
      return '<a href="'+url+'">'+url+'</a>';
    }).replace(/\B@([_a-z0-9]+)/ig, function(reply) {
      return reply.charAt(0)+'<a href="http://twitter.com/
'+reply.substring(1)+'">'+reply.substring(1)+'</a>';
    });
    statusHTML.push('<li><span>'+status+'</span>
    <a style="font-size:85%" href="#" onclick=
    NKOpenURLInSafari("http://twitter.com/'+username+
    '/statuses/'+
twitters[i].id+'")>'+relative_time(twitters[i].
created_at)+
    '</a></li>');
  }
  document.getElementById('twitter_update_list').innerHTML
  = statusHTML.join('');
}


function relative_time(time_value) {
  var values = time_value.split(" ");
  time_value = values[1] + " " + values[2] + ", " +
  values[5] + " " + values[3];
  var parsed_date = Date.parse(time_value);
  var relative_to = (arguments.length > 1) ? arguments[1] :
  new Date();
  var delta = parseInt((relative_to.getTime() - parsed_
  date) / 1000);
  delta = delta + (relative_to.getTimezoneOffset() * 60);
```

```
if (delta < 60) {
  return 'less than a minute ago';
} else if(delta < 120) {
  return 'about a minute ago';
} else if(delta < (60*60)) {
  return (parseInt(delta / 60)).toString() + ' minutes
  ago';
} else if(delta < (120*60)) {
  return 'about an hour ago';
} else if(delta < (24*60*60)) {
  return 'about ' + (parseInt(delta / 3600)).toString() +
  ' hours ago';
} else if(delta < (48*60*60)) {
  return '1 day ago';
} else {
  return (parseInt(delta / 86400)).toString() + ' days
  ago';
}
}
```

Keep the file name for reference purposes and name it blogger.js in your project.

NOTE Making sure your .js file is in the app bundle Sometimes when you add your own .js file to your Xcode project, the file is not automatically included in the app bundle. To fix Xcode's oversight, simply drag the .js file to Targets > MyApp > Copy Bundle Resources in the Groups & Files pane. This copies the file to the app bundle to ensure that everything works smoothly.

With this script in your toolbox, you can write a brief amount of HTML that will pull tweets in from a specified account. The portion that performs the Twitter magic is highlighted:

```
<html>
<head>
<meta name = "viewport" content = "initial-scale = 1.0,
user-scalable = no" />
<link href="style.css" rel="stylesheet" type="text/css">

<script type="text/javascript" src="NKit.js"></script>
<script type="text/javascript">

var navController = new NKNavigationController();
navController.setTitle("Klayon's Tweets");
navController.setTintColor(255, 0, 0);

// check for internet connectivity
function checkForInternet()
{
```

```
  var isInternetAvailable = false;
  if (NKIsInternetAvailableViaWifi()==1)
  {
    isInternetAvailable = true;
  }
  else
  {
    if (NKIsInternetAvailableViaCellularNetwork()==1)
    {
    NKAlert("Info", "Internet is available only via
    cellular network,
    carrier fees may apply.");
    isInternetAvailable = true;
    }
  else
    {
    NKAlert("Error", "An internet connection is required to
    access content
    for this function.");
    isInternetAvailable = false;
    }
  }
  return isInternetAvailable;
}
if (checkForInternet())
{
  // load content from the web
}

</script>
</head>

<body>

<!--
This unordered list displays the tweets,
and pulls in as many as you specify in count=n
in the JavaScript below.
-->

<ul id="twitter_update_list">
</ul>
```

```html
<script type="text/javascript" src="blogger.js"></script>
<script type="text/javascript" src="http://twitter.com/
statuses/
  user_timeline/klayon.json?callback=twitterCallback2
  &amp;count=10"></script>

</body>
</html>
```

Finally, to give this unordered list some minimal style for readability, declare a few CSS rules for this view in a file called style.css:

```css
body {
  margin: 0px;
  padding: 0px;
  font-family: 'Times New Roman';
}

ul {
  list-style-type: none;
}

#twitter_update_list {
  font-size: 14pt;
  list-style-type: none;
  margin: 0px;
  padding: .5em;
}

#twitter_update_list li {
  margin: 0px;
  padding: .5em;
  border-bottom-width: 1px;
  border-bottom-style: solid;
  border-bottom-color: #666;
}
```

Combining the blogger.js, main.html, and style.css files gives you a page like that shown in **Figure 5.6**, using my own Twitter account as an example.

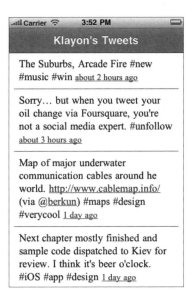

5.6 Bringing a Twitter feed into a NimbleKit app.

Using Twitter as a light app content management system in this way is pretty cool—it's a convenient way to update people on the go about news or specials, just as organizations and businesses use Twitter to update people at their desktops. And while there are Twitter iOS apps that allow people to do this already, nothing says you love your clients or customers more than a very targeted and custom-designed experience. You can style this however you want, not to mention make it one of several other content screens in an app.

Working with images

We've talked about some ways of working with text content. Working with images in NimbleKit is also fairly simple. There are two main methods to display images: within an app's content (inline), and as an image overlay.

Images in content

Placing images inline is just as easy as doing it in a web page:

```
<img src="filename.jpg" width="x" height="y"
class="classname">
```

And by assigning an image a CSS class, you're empowered to style it as you wish.

Let's revisit the example of the Bill of Rights structured as a definition list. Wouldn't it look better with a photo of a bald eagle above it? Sure it would!

And it just so happens that I have a nice JPG of an eagle to add, sized for the iPhone (**Figure 5.7**).

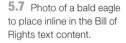

5.7 Photo of a bald eagle to place inline in the Bill of Rights text content.

Let's add it to the HTML, right above the text:

```
<img src="eagle.jpg" width="280" height="206">
<h1>United States Bill of Rights</h1>
<dl>
<dt>1<sup>st</sup> Amendment</dt>
<dd>Congress shall make no law respecting an establishment
of religion, or prohibiting the free exercise thereof; or
abridging the freedom of speech, or of the press; or the
right of the people peaceably to assemble, and to petition
the Government for a redress of grievances.</dd>
```

```
<dt>2<sup>nd</sup> Amendment</dt>
<dd>A well regulated Militia, being necessary to the
security of a free State, the right of the people to keep
and bear Arms, shall not be infringed.</dd>...
</dl>
```

Figure 5.8 shows the results.

5.8 The Bill of Rights definition list with the eagle image added.

This works predictably well but could certainly use some more styling. In fact, the image runs right up to the left, top, and right edges of the screen—and that looks fine—but the text content could use a margin. As long as we're considering that, why not add a margin around the photo? And as long as we're doing that, why not try out some CSS3 and see whether adding rounded corners finishes it off with a bit more style?

To do this, let's start by proportionally resizing the image a bit smaller so there's a margin of 20 pixels around it:

```
<img src="eagle.jpg" width="280" height="206">
```

To match this, add 20 pixels of margin to the body. This can be done with CSS:

```
body {margin: 20px;}
```

Finally, let's add some CSS3 curviness with border-radius; with NimbleKit, you don't even need to use the WebKit version. It understands the straight CSS3 style just fine:

```
img {border-radius: 15px;}
```

Adding a margin and rounded corners creates a more finished product (**Figure 5.9**).

5.9 The Bill of Rights definition list with the image added, now featuring margins and rounded corners!

I hope this demonstrates some of the exciting design possibilities available to you as a web designer using NimbleKit with CSS3. (We'll explore more CSS3 fun later in Chapter 9.)

Image overlays

Images can be more than just inline content in iOS apps. NimbleKit also supports a method of placing an image on top of a content screen called **NKImageView**. Here's how it works.

Similar to using **NKTableView**, **NKImageView** is specified with JavaScript in the head of an HTML file. First, we need to create a new variable and call on the **NKImageView** control.

Let's see how this would work in a hypothetical iOS app that teaches people about outdoor wilderness areas. **Figure 5.10** shows a text screen about the Boundary Waters Canoe Area Wilderness in northeastern Minnesota, one of my favorite places to go on vacation.

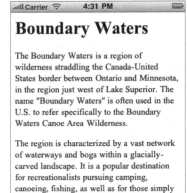

5.10 Text about the Boundary Waters (paraphrased from Wikipedia).

Here's the HTML file that, when added to a new NimbleKit-based Xcode file, creates the screen shown in Figure 5.10:

```
<html>
<head>
<meta name = "viewport" content = "initial-scale = 1.0,
user-scalable = no" />
<script type="text/javascript" src="NKit.js"></script>
</head>
<body>
<h1>Boundary Waters</h1>
<p>The Boundary Waters is a region of wilderness
straddling the Canada-United States border between Ontario
and Minnesota, in the region just west of Lake Superior.
The name "Boundary Waters" is often used in the U.S.
to refer specifically to the Boundary Waters Canoe Area
Wilderness.</p>
```

```
<p>The region is characterized by a vast network of
waterways and bogs within a glacially carved landscape.
It is a popular destination for recreationalists pursuing
camping, canoeing, and fishing, as well as for those simply
looking for natural scenery and relaxation.</p>
<p>Source &mdash; http://.en.wikipedia.org</p>
</body>
</html>
```

This is a good start, but it's pretty boring. We need to add some style, as well as the **NKImageView** control so the app's users can see an image of the Boundary Waters. First, the **NKImageView**—let's give this instance the name **photo**:

```
var photo = new NKImageView;
```

Next, create a toolbar at the bottom of the screen. It will contain a button for showing the photo:

```
var toolBar = new NKToolBar();
toolBar.init(416);
toolBar.addButton("Show photo", "", "");
toolBar.setStyle("blacktranslucent");
toolBar.show();
```

There are several things going on here:

1. The first line of code initializes the NKToolbar control and declares a variable name for it—this is called toolBar.

2. The next line positions the toolbar on the screen on the y-axis. If you start with the full height of the screen (480 pixels), subtract 20 pixels for the status bar, and then subtract 44 pixels for the height of the toolbar, the upper edge of the toolbar is placed at 416.

3. The next line adds a button labeled "Show photo."

4. The last two lines make the toolbar translucent black and display it on screen.

Now we need an image to add to the **NKImageView**. **Figure 5.11** shows a photo I took during one of my trips to the Boundary Waters Canoe Area Wilderness. I've added a frame to make it pop a bit more on screen, and a "CLOSE X" notation in the corner so users are prompted to click on the image to make it go away.

5.11 Photo of people canoeing in the Boundary Waters.

Next, we need to initialize the control with position, size, and the name of the image. Place the corner of the JPG at coordinate 44, 20 so it's centered in the screen:

```
photo.init (44, 24, 233, 367, 'bwca.jpg');
```

We're getting close; next we need to write a callback so the button we added to the toolbar will display the image. Here's how that is coded:

```
toolBar.addButton("Show photo", "", "photo.show()");
```

Finally, one more useful detail for **NKImageView** is that tapping the image can also hide it (and remember, I've prepped the image in Photoshop to indicate this already). Add this line to the code at the end of the **photo** variable:

```
photo.setOnClickCallback("photo.hide()");
```

The finished code should look like this:

```
<script type="text/javascript">

var photo = new NKImageView;
photo.init (44, 24, 233, 367, 'bwca.jpg');
photo.setOnClickCallback("photo.hide()");

var toolBar = new NKToolBar();
toolBar.init(416);
toolBar.addButton("Show photo", "", "photo.show()");
toolBar.setStyle("blacktranslucent");
toolBar.show();

</script>
```

The main.html file should now look like this:

```
<html>
<head>
<meta name = "viewport" content = "initial-scale = 1.0,
user-scalable = no" />
<script type="text/javascript" src="NKit.js"></script>
<script type="text/javascript">

var photo = new NKImageView;
photo.init (44, 24, 233, 367, 'bwca.jpg');
photo.setOnClickCallback("photo.hide()");

var toolBar = new NKToolBar();
toolBar.init(416);
toolBar.addButton("Show photo", "", "photo.show()");
toolBar.setStyle("blacktranslucent");
toolBar.show();

</script>
</head>
<body>

<h1>Boundary Waters</h1>
<p>The Boundary Waters is a region of wilderness
straddling the Canada-United States border between Ontario
and Minnesota, in the region just west of Lake Superior.
The name "Boundary Waters" is often used in the U.S.
```

```
to refer specifically to the Boundary Waters Canoe Area
Wilderness.</p>
<p>The region is characterized by a vast network of
waterways and bogs within a glacially carved landscape.
It is a popular destination for recreationalists pursuing
camping, canoeing, fishing, as well as for those simply
looking for natural scenery and relaxation.</p>
<p>Source &mdash; http://.en.wikipedia.org</p>
</body>
</html>
```

This should all work pretty well now—except that the text is not styled at all. Add a style.css file to the app bundle and style the content a little, making it a different typeface and adding background color for a bit of contrast:

```
body {
  background-color: #f1f1e6;
  font-family: "Trebuchet MS";
  font-size: .95em;
  margin: 20px;
  padding: 0px;
}
```

After adding `<link href="style.css" rel="stylesheet" type="text/css" />` to the head of the HTML file, the results will look like **Figure 5.12** before the button is pressed to display the image.

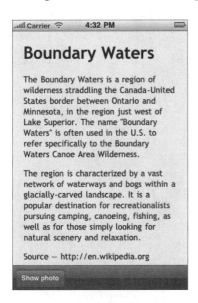

5.12 The styled Boundary Waters text content.

And after the button is pressed, the screen looks like **Figure 5.13**.

5.13 The Boundary Waters photo overlaying the text content.

NKImageView has some potential as a great little control, especially when paired with a toolbar and button to show the image. But I'm sure you can think of additional creative ways to use this aside from the example described here!

Summary

Working with text and image content is essential for any iOS app, and this chapter scratched the surface of the many possibilities for incorporating and styling content. In this chapter you learned how to

- Structure HTML content several different ways using definition lists.
- Add social content, live from the web, to an app; in our example, we used a feed from Twitter.
- Add an image inline right into app content.
- Add an image overlay on top of app content.

This introduced you to a few techniques for exploring how you might design custom content-based apps for use offline or online. Next, we'll look at designing specifically for people who are mobile: Chapter 6 is about geo-location on iOS devices, and looks at a few ways to design a Google Maps view into a NimbleKit-based app.

6 FOCUS ON APP CONTENT: MAPS

iOS apps are designed to run
natively on Apple mobile devices.

Maps help mobile natives know
where they are running.

Mapping is one of my favorite uses of my iPhone, and it's an example of a good philosophy for mobile application design. It fits squarely into my approach about why designing custom, native apps can offer real value to people by providing a useful solution:

- Mapping is a timeless need that can surface wherever we are and regardless of what we are doing. It is both "real world" and technology agnostic—though, in this case, mobile technology can have a role as a mapping solution!

- Mapping is a generic activity that applies to any business or organization that has a geographic, bricks-and-mortar location. So it can be a key feature of a mobile app for nearly anyone.

- Mapping is about more than finding the location of a business, organization, or other venue. On a cellular iOS device, it can also be a wayfinding tool that provides detailed driving, mass transit, or walking directions from your current location to your destination.

And once you start seeing mapping as a universally appreciated function that can apply to nearly any client or employer, you start seeing all kinds of applications. Mapping is really the location of

- restaurants
- stores
- event centers
- festivals
- farmers' markets
- medical clinics
- campgrounds
- people
- services

Clearly, this list could go on and on. But again, part of why I'm writing this book is to think aloud as a designer, and inspire you to see opportunities for designing apps that are relevant to clients and, more importantly, relevant to their customers. Mapping, then, becomes a tool that anyone should be able to relate to when discussing an iOS native app project.

The NimbleKit framework offers a handy map API that pulls in the same Google Maps view that you get in the actual Maps application that comes on the iPhone and iPad. Right out of the box, it's fairly easy to configure, and has only one minor shortcoming at the moment, which has a relatively easy work-around.

I'll be up-front and start with the shortcoming: Apple does not currently open up the full Google Maps API to app designers. So we can pull a live map view into our custom app and also set a pin and label it to annotate our map, but we cannot currently pull the directions feature into our app. I suspect this is part of Google's agreement with Apple for being the default mapping tool on the iOS platform. (And we may not like it, but it is what it is.)

Fortunately, there is a nice work-around to this shortcoming: providing a link to the Maps app within the map view. In other words, even though we cannot fully incorporate directions into our own apps, we can still link to a map pin in Google Maps and have it open directly in the Maps application. It adds an additional step, but once you see how seamlessly it works, the extra step ends up feeling rather trivial.

So let's walk through a few ways that we can pull a custom Google map view into a native app, and still have it lead to getting convenient directions from our current location. We can design all of this with a dash of HTML, a few lines of CSS, a little JavaScript, and some help from NimbleKit's Objective-C framework.

Method one: Using NKButton

This method is pretty short and sweet. It allows us to pull in a Google Maps view, specify a pin at a particular location and label it, then add a "View in Google Maps" `NKButton` with a native appearance and behavior that will link to the same location in Maps so users can get driving directions. And it does this all without requiring you to do any styling or even much coding. In fact, it hardly even uses any HTML; it's all NimbleKit and a bit of JavaScript at the beginning of the file.

I'm including this technique as well as a more custom-designed button because I want to highlight how the NimbleKit Objective-C framework often supports both native features and custom features. Learning how to do both is very useful: Sometimes the specification of an app might call for a native look, and other times your employer's or client's branding guidelines might lead you toward a less native, more customized styling of an element like a button by using CSS.

Planning the screen layout

This technique allows you to overlay a button on the screen contents, so we'll do that in a single screen. And for this example we'll dispense with any additional elements except the status bar and title bar, which are present in any screen of a content-based app.

Looking at our 320 × 480 pixel iPhone screen, here are the elements we're starting with:

- status bar: 320 × 20 pixels
- title bar: 320 × 44 pixels

That leaves us with 416 pixels in height to work with; this comes into play in the JavaScript step of implementing this technique.

Writing the HTML

The HTML for this method is probably the shortest ever written. In fact, as a stand-alone page of code, it really doesn't do anything until the JavaScript is added later:

```
<html>
<head>
<meta name = "viewport" content = "initial-scale = 1.0,
  user-scalable = no" />
</head>
<body>
</body>
</html>
```

Truly, that's it. We're just creating a shell in which to place some JavaScript, and setting the screen to not be scalable with touch gestures with the `meta` tag (the map will be scalable already). Thus, this method requires no CSS, as it's not necessary for using NKButton.

Writing the JavaScript

Now we get to the interesting part: calling the NimbleKit framework into action for the embedded map view. Time to dust off some JavaScript writing skills (but it's pretty easy).

There are four basic styles and behaviors that we call into action in this step:

1. Give the page a name in the native iPhone UI title bar.

2. Set the color of the title bar.

3. Embed the Google map view.

4. Enable a NimbleKit button to link to our location in Maps.

The first two steps were already detailed in Chapter 4, so here's what I've done to name the screen and set the color to red:

```
var navController = new NKNavigationController();
navController.setTitle("Our Map Page");
navController.setTintColor(255, 0, 0);
```

Now let's jump to the parameters that are available in the map embed code. The first task is to create a new **NKMapView**, which is a variable with the name "map":

```
var map = new NKMapView();
```

After this, define the size of the view by specifying the X and Y coordinates of its point of origin (with 0, 0 being the top left of the screen) and decide how wide and tall it should be. So, a map taking up an entire iPhone screen would have these values:

```
map.init(0, 0, 320, 480);
```

But remember that you have two other elements on the screen (the status bar and the title bar), so you need to subtract those heights from the available height to calculate the remainder:

```
map.init(0, 0, 320, 416);
```

Then set the map type. As in Apple's Maps app, there are three settings for the type of map: **standard**, **satellite**, and **hybrid**. In this example, it's set to **"standard"**:

```
map.setMapType("standard")
```

The next part is a bit tricky and requires some experimentation: Set the geographic center of the map view and set the range that the view is covering. The format is

```
map.setDisplayRegion(latitude, longtitude, latitudeDelta,
longtitudeDelta);
```

Latitude and longitude are measured from the intersection of the prime meridian and the equator, so values of latitude north of the equator are positive (and south are negative), and values of longitude east of the prime meridian are positive (and west are negative). The trickier part is setting the delta values. Hopefully my experimentation here is helpful to you: The value of .01 is equal to about 5,500 feet, or approximately one mile.

In the following code example, I'll use a restaurant called Sven and Ole's and their real location in the town of Grand Marais, Minnesota (pay them a visit if you're ever in northern Minnesota—they serve great pizza!):

```
map.setDisplayRegion(47.7494, -90.3336, .01, .01);
```

NOTE Getting longitude and latitude in Google Maps
Finding longitude and latitude in Google Maps is easy if you use the right tool. I use the LatLng Tooltip (by Marcelo C). Find it by going to Google Maps Labs and enable LatLng Tooltip. You can find this and other Google Maps Labs options by clicking on the New! (with a green lab beaker icon) link in the upper right corner of the Google Maps screen, next to Help and Sign in.

The next line tells the map view not to show the user's current location. In this example, that's not important, nor is it necessarily possible if the user is outside the range of the map view:

```
map.showUserLocation("no");
```

Next, I'll set a pin at the location of Sven and Ole's restaurant. It makes sense to me to initially center the map view on the restaurant, so you'll see that the latitude and longitude match the `setDisplayRegion` parameters. Then come two labels or titles—the first being the main label for the pin and the second being an optional subtitle:

```
map.addAnnotation(47.7494, -90.3336, "Sven and Ole's",
"Restaurant");
```

NOTE A word about NimbleKit's MapView

`NKMapView` supports some additional parameters following the title and subtitle, including image, color, and an optional callback function. But I want to keep my pin like the ones I'm accustomed to seeing in the actual iOS Maps application, so I recommend keeping your map pin simple as in my example. If you want to explore these options, consult the NimbleKit documentation for `NKMapView`, which explains them in more detail.[1]

At this point, I'll add a line of code to set a slight delay before the pin appears (to match the user experience in Maps):

```
setTimeout(selectAnnotation, 500);
```

I'll also add a few brief lines to make the map function actually display the map and allow the pin to be selectable:

```
function selectAnnotation()
{
  map.selectAnnotation('Here');
}
```

Next, I'll define the function to navigate to the same location in Maps. I've called it `openInMaps` and in it I'm placing a URL for Sven & Ole's location in Google Maps:

```
function openInMaps()
{
NKOpenURLInSafari('http://maps.google.com/?q=Sven+%26+Ole%
27s&cid=11769511904778742301');
}
```

ALL GOOGLE MAPS LINKS ARE NOT EQUAL!

One very important note here about the origin of the Google Maps link: Create it on your iOS device and then email it to yourself. One of the more interesting debugging exercises I went through when first implementing an **NKMapView** was having some frustrating things happen with this link, such as

- The link opening in Safari instead of Maps
- And, worse yet, sometimes not even having the right location open

Some of this can be chalked up to the occasional Google Maps bug; I've sent myself map links out of Maps before and sometimes the link in the email doesn't go back to the actual place I was seeing in Maps. I had this experience way before I became an app designer, so I wasn't totally surprised to have it happen when I tried to embed a map into an app.

But I also learned something valuable along the way: There's more than one format for a Google Maps link! Sure enough, it's one of those things that you don't necessarily pay much attention to until you run into a bug—at least, that's when I first noticed. And once I did, some additional experimentation led me to realize that the iPhone will not necessarily open a Google Maps link in the Maps app unless the link was generated by Maps itself. This is why you should always generate your map links on your iOS device. Then it should open properly in Maps rather than Safari.

The last step is implementing **NKButton** so that someone using this app screen can navigate to the same location in the Maps application. The following code defines a variable **mapsButton**, sets the size of the button, and calls on the **openInMaps** function when clicked. I'll also set the alpha (transparency) to 80 percent so that it has a nice semitransparent appearance:

```
var mapsButton = new NKButton();
mapsButton.init(70, 10, 180, 40, "openInMaps");
mapsButton.setTitle("View in Google Maps");
mapsButton.show();
NKNativeControlSetAlpha(mapsButton, 0.8);
```

The real last step, checking for Internet connectivity and using **NKAlert**, was covered in Chapter 5; the code is the same as it was in that example.

Taking all these pieces together, the HTML with JavaScript now looks like this, with the NimbleKit sections highlighted:

```html
<html>

<head>
<meta name = "viewport" content = "initial-scale = 1.0,
  user-scalable = no" />

<script type="text/javascript" src="NKit.js"></script>
<script type="text/javascript">

var navController = new NKNavigationController();
navController.setTitle("Our Map Page");
navController.setTintColor(255, 0, 0);

var map = new NKMapView();
map.init(0, 0, 320, 416);
map.setMapType("standard");
map.setDisplayRegion(47.7494, -90.3336, .01, .01);
map.showUserLocation("no");
map.addAnnotation(47.7494, -90.3336, "Sven and Ole's",
"Restaurant");
map.show();
setTimeout(selectAnnotation, 500); //wait for half second

function selectAnnotation()
{
  map.selectAnnotation('Here');
}

function openInMaps()
{
NKOpenURLInSafari('http://maps.google.com/?q=Sven+%26+Ole%
27s&cid=11769511904778742301');
}

var mapsButton = new NKButton();
mapsButton.init(70, 10, 180, 40, "openInMaps");
mapsButton.setTitle("View in Google Maps");
mapsButton.show();
NKNativeControlSetAlpha(mapsButton, 0.8);
```

```
// check for internet connectivity

function checkForInternet()
{
  var isInternetAvailable = false;
  if (NKIsInternetAvailableViaWifi()==1)
  {
    isInternetAvailable = true;
  }
  else
  {
    if (NKIsInternetAvailableViaCellularNetwork()==1)
    {
    NKAlert("Info", "Internet is available only via
cellular network,
    carrier fees may apply.");
    isInternetAvailable = true;
    }
  else
    {
    NKAlert("Error", "An internet connection is required to
access content
    for this function.");
    isInternetAvailable = false;
    }
  }
  return isInternetAvailable;
}

if (checkForInternet())
{
  // load content from the web
}

</script>
</head>

<body>
</body>

</html>
```

When tested on an iPhone, the results of this work look like what's shown in **Figure 6.1**.

6.1 Our app screen using **NKMapView** with **NKButton**.

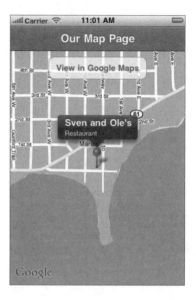

You can do this exercise and still partially test it in Simulator, but because Simulator doesn't have the Maps app on it, the button causes Google Maps to open in Safari instead. Full testing requires you to provision and test on an iPhone, iPod touch, and iPad.

Method two: Styling an HTML button

This next method uses a CSS-styled HTML button instead of **NKButton**. My example is only one way to style a button with CSS: The idea here is that with styles (and especially with additional elements like graphics files), you can design a more branded appearance with this detail.

Planning the screen layout

In this map example, I'll continue using the status bar, title bar, and tab bar page layout from our previous examples. The reason for this is practical: A map screen in an app will probably be one of multiple screens. Because

I like my examples to be practical and reusable for actual applications, I'll move ahead with the map exercise using this assumption.

Looking at the 320 × 480 pixel iPhone screen, start with the same elements as we did before:

- status bar: 320 × 20 pixels
- title bar: 320 × 44 pixels

This leaves us with 416 pixels of height to work with again. But this time we'll use some of that space for the HTML and CSS button as it cannot overlay the map like `NKButton` does. I'll propose adding a button that's 32 pixels tall below the map and giving it 50 pixels of margin on the left and right to center it, and 5 pixels of margin on top so it doesn't directly abut the map. I'll also style the space around the button to look similar to a native tab bar (dark gray).

With this as our plan, let's proceed with building it out!

Writing the HTML

The HTML for this is remarkably brief, as most of this code example utilizes NimbleKit's map architecture. About the only things we're doing in the HTML are setting the stage for getting things wired together and creating the button that links to Maps.

To get this first step done, create a new HTML file. As with the last method, the initial code will scale the viewport to the iPhone and set user scalability to `no`. The embedded map will already be scalable, but we don't want the button to be scalable as this would result in a nonnative user experience. We'll also include a few lines to define a class named `"button"` and build it as an ordered list item (as you typically would when designing a navigational element in a website). So, here's what this first step looks like:

```
<html>

<head>
<meta name = "viewport" content = "initial-scale = 1.0,
user-scalable = no" />
</head>
```

```
<body>
<ul class="button">
<li><a href = "http://maps.google.com/?q=Sven+%26+Ole%27s&
cid=11769511904778742301">View in Google Maps</a></li>
</ul>
</body>

</html>
```

Writing the CSS

Next, style the button in a native way, giving it the native iOS UI treatment of embossed text, rounded corners, drop shadow, and so on. Here's the CSS for this, using the specs we decided on in the Planning the Screen Layout section—put this in a file called style.css. I had to work one step ahead to get my top margin value for the button; I needed to calculate the height of the map view (366 pixels) to arrive at the top margin value (366 + 5 = 371 pixels).

```
body {
  margin: 0px;
  padding: 0px;
  font-family: 'Helvetica';
  background-color: #333;
}

.button {
  font-size: 14pt;
  display: inline;
}

ul.button {
  list-style-type: none;
}

ul.button li a {
  color: #fff;
  background-color: #1a5c00;
  width: 220px;
  text-align: center;
  padding: 8px 0px 0px 0px;
```

And in place of the `NKButton` JavaScript, there's a small `onClick` event doing the work instead, which changes the formatting of the link like so:

```
<a href="#" onclick="NKOpenURLInSafari('http://maps.
google.com/?q=Sven+%26+Ole%27s&cid=11769511904778742301')"
>View in Google Maps</a>
```

After inserting the JavaScript into the HTML file, this is what you have:

```
<html>
<head>
<meta name = "viewport" content = "initial-scale = 1.0,
user-scalable = no" />
<link href="style.css" rel="stylesheet" type="text/css">

<script type="text/javascript" src="NKit.js"></script>
<script type="text/javascript">

var navController = new NKNavigationController();
navController.setTitle("Where are Sven and Ole?");
navController.setTintColor(255, 0, 0);

var map = new NKMapView();
map.init(0, 0, 320, 366);
map.setMapType("standard");
map.setDisplayRegion(47.7494, -90.3336, .01, .01);
map.showUserLocation("NO");
map.addAnnotation(47.7494, -90.3336, "Sven and Ole's",
"Restaurant");
map.show();
setTimeout(selectAnnotation, 500); //wait for half second
function selectAnnotation()
{
  map.selectAnnotation('Here');
}


// check for internet connectivity

function checkForInternet()
{
  var isInternetAvailable = false;
  if (NKIsInternetAvailableViaWifi()==1)
```

```
  {
    isInternetAvailable = true;
  }
  else
  {
    if (NKIsInternetAvailableViaCellularNetwork()==1)
    {
    NKAlert("Info", "Internet is available only via
cellular network,
    carrier fees may apply.");
    isInternetAvailable = true;
    }
  else
    {
    NKAlert("Error", "An internet connection is required to
access content
    for this function.");
    isInternetAvailable = false;
    }
  }
  return isInternetAvailable;
}

if (checkForInternet())
{
  // load content from the web
}

</script>

</head>
<body>

<ul class="button">
<li><a href="#" onclick="NKOpenURLInSafari('http://maps.
google.com/?q=Sven+%26+Ole%27s&cid=11769511904778742301')"
>View in Google Maps</a></li>
</ul>

</body>
</html>
```

The results look like what is shown in **Figure 6.2**.

6.2 The embedded Google map view with a link to Maps, this time styled with a custom button.

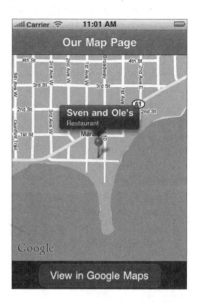

iPad considerations

This last example targets the iPad, so it primarily changes the math involved in planning the screen real estate. But I'm also using it as an opportunity to change the design slightly. To give this example a try, open a new NimbleKit project, and choose iPad as the device.

Planning the screen layout

Starting with the 768 × 1024 pixel iPad screen, let's begin with the same UI elements we used in the previous two examples—they're just wider this time:

- status bar: 768 × 20 pixels
- title bar: 768 × 44 pixels

This leaves us with 960 pixels of height to work with on the iPad. That's a lot of space, and typically we're seeing a bit more "app chrome" (styled areas that frame content or navigation) in iPad designs. So let's set a border area of 30 pixels around the map view and a bit more spacing around the button, too. Doing the math leaves us with a map view height of 850 pixels.

There's nothing new in the HTML itself, so let's jump ahead to the CSS where I've made a few changes.

Writing the CSS

Following through on my plan to have a bit of chrome around the iPad map view, this example has the style.css file call up a metal texture background image for the body and button, giving it an embossed metal look (you can download this GIF file from iosapps.tumblr.com; look under Table of Contents > Chapter 6). Once again, I had to work one step ahead to get the top margin value for the button. I needed to calculate the height of the map view (850 pixels) in order to arrive at the top margin value: 850 pixels + 30 pixel margin above the map + 20 pixels between the map and button = 900 pixels. So here's the CSS for this version—note how the button margins also had to change because of the larger iPad screen:

```
body {
  margin: 0px;
  padding: 0px;
  font-family: 'Helvetica';
  background-color: #999;
  background-image: url(metalTexture.gif);
  background-position: 100% 100%;
}

.button {
  font-size: 14pt;
  display: inline;
}

ul.button {
  list-style-type: none;
}

ul.button li a {
  color: #666;
  background-color: #999;
  width: 220px;
```

```css
    text-align: center;
    padding: 8px 0px 0px 0px;
    margin: 900px 274px 0px 274px;
    height: 32px;
    border: 2px ridge #ccc;
    display: block;
    border-radius: 10px;
    text-shadow: 0px 2px 0px #ccc;
    text-decoration: none;
    background-image: url(metalTexture.gif);
}
```

Writing the JavaScript

The JavaScript for this method is nearly identical to that in the previous method, so I'll only highlight the difference: the larger specification for the map view. The two values of 30 are the X and Y coordinates for where the iPad starts drawing the map view, establishing the corner and border on the upper left; 708 and 850 pixels are the width and height respectively, so you have a 30-pixel right margin plus plenty of extra space under the map view for the button and extra chrome area:

```javascript
map.init(30, 30, 708, 850);
```

The resulting HTML file, with JavaScript, for the iPad is

```html
<html>
<head>
<meta name = "viewport" content = "initial-scale = 1.0,
user-scalable = no" />
<link href="style.css" rel="stylesheet" type="text/css" />

<script type="text/javascript" src="NKit.js"></script>
<script type="text/javascript">

var navController = new NKNavigationController();
navController.setTitle("Where are Sven and Ole?");
navController.setTintColor(255, 0, 0);

var map = new NKMapView();
map.init(30, 30, 708, 850);
map.setMapType("standard");
```

```
map.setDisplayRegion(47.7494, -90.3336, .01, .01);
map.showUserLocation("NO");
map.addAnnotation(47.7494, -90.3336, "Sven and Ole's",
"Restaurant");
map.show();
setTimeout(selectAnnotation, 500); //wait for half second
function selectAnnotation()
{
  map.selectAnnotation('Here');
}


// check for internet connectivity

function checkForInternet()
{
  var isInternetAvailable = false;
  if (NKIsInternetAvailableViaWifi()==1)
  {
    isInternetAvailable = true;
  }
  else
  {
    if (NKIsInternetAvailableViaCellularNetwork()==1)
    {
    NKAlert("Info", "Internet is available only via
cellular network,
    carrier fees may apply.");
    isInternetAvailable = true;
    }
  else
    {
    NKAlert("Error", "An internet connection is required to
access content
    for this function.");
    isInternetAvailable = false;
    }
  }
  return isInternetAvailable;
}


if (checkForInternet())
{
```

```
    margin: 371px 50px 0px 50px;
    height: 32px;
    border: 1px solid #666;
    display: block;
    border-radius: 10px;
    text-shadow: 0px 1px 0px #ccc;
    text-decoration: none;
}
```

Note some of the details that still make this button look native: 14-point text, the CSS3 border-radius effect, and the text-shadow all contribute to making this button fit into the iOS user interface rather smoothly. It's just that in this case, I wanted a green button.

Of course, the point of designing a button with CSS might be to do something much more adventurous than making an otherwise native-looking button green. By making the map view narrower (with a margin surrounding it) and visually integrating a button into this chrome, you could design the button to look like nearly anything you wish. For example, you might put the map in a woodgrain frame and use a Photoshop filter to carve the text into a woodgrain background.

Anyway, I digress: After this you can experiment with button styles as much as you want because the sky is the limit. To wrap up the styling for this exercise, save the file and add a line to your HTML to call up the styles. Let's call it style.css:

```
<link href="style.css" rel="stylesheet" type="text/css">
```

If you run this at this point, you'll see your button but it will not properly link to Maps yet, because you haven't added the NimbleKit code to support it.

Writing the JavaScript

The JavaScript for this method is very similar to that of the previous method, so I'll only highlight the differences. First, remember the height value for the NKMapView that I calculated earlier for the CSS styling. I'm now putting that 366-pixel value into the function for the y-height:

```
map.init(0, 0, 320, 366);
```

```
  // load content from the web
}

</script>

</head>
<body>

<ul class="button">
<li><a href="#" onclick="NKOpenURLInSafari('http://maps.
google.com/?q=Sven+%26+Ole%27s&cid=11769511904778742301')"
>View in Google Maps</a></li>
</ul>

</body>
</html>
```

The results are shown in **Figure 6.3**.

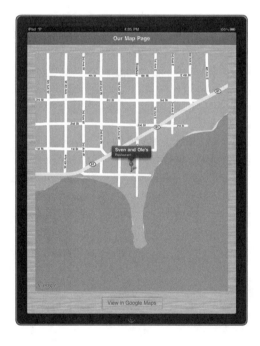

6.3 The app screen using NKMapView on the iPad.

Summary

In this chapter you learned how to

- Pull in an **NKMapView** that displays a Google map view and supports a location pin with title and subtitle.
- Add an **NKButton** for linking the in-app map view to the same location in Maps, so users can get driving directions to their destination from their current location.
- Design a custom Maps button using HTML and CSS.
- Design the NKMapView scaled to the iPad, and in a way that allows for a custom chrome around the map view.

Your focus on iOS app content began with text and images and has continued with map and location content. Next you will see how your web standards skills, coupled with NimbleKit, allow you to include audio and video content in native iOS apps.

REFERENCES

1. http://www.nimblekit.com/

7 FOCUS ON APP CONTENT: AUDIO

In addition to text and images, social content via feeds, and mapping, the NimbleKit Objective-C framework supports rich media content such as audio and video.

While not every web designer is an audio engineer, today's Macs can make basic audio production pretty accessible. For example, after importing a song into GarageBand, you can make some fairly quick work out of trimming it and perhaps adding fades. Doing so can give you a brief song sample, like what you can listen to when shopping iTunes for music.

This process makes designing an iOS app with an audio introduction, song sample, or even spoken directions possible. A custom iOS app for education, a musical group, or a product can be greatly enriched with audio content.

NimbleKit supports playing audio content in iOS apps in multiple ways. Two of them are demonstrated in this chapter: HTML5 and NimbleKit's NKAudioPlayer.

Playing audio with HTML5

Playing an audio file with HTML5 could not be any easier. You need only use HTML5's new `audio` tag and have an MP3 audio file handy. For example, to play a 30-second song sample called song.mp3, this is all that's needed:

```
<audio src="song.mp3" controls></audio>
```

The addition of `controls` is necessary; without it, there's no way to cue the track to play.

So at its most basic, playing this MP3 file in NimbleKit requires the following HTML:

```
<html>
<head>
<meta name = "viewport" content = "initial-scale = 1.0,
user-scalable = no">
<script type="text/javascript" src="NKit.js"></script>
</head>

<body>
<audio src="song.mp3" controls></audio>
</body>

</html>
```

The result in Simulator looks like what's shown in **Figure 7.1**.

7.1 The native iOS audio player that is displayed using the HTML5 audio element.

So how can we make this a bit more interesting, you might be wondering? Well, if I were checking out a song sample in an iOS app, I might want to see who the musician is, so let's include an image (photo.jpg). Additionally, the positioning and sizing of the controls can be set by a stylesheet so it is centered low in the screen, overlaying the photo. Here are a few rules for this example's style.css:

```css
body {
  margin: 0px;
  padding: 0px;
}

audio {
  width: 100%;
  position: absolute;
  z-index: 10;
  top: 390px;
}
```

You can also add a title bar. In this example, the title bar background is coordinated with the photo. Here's the revised HTML:

```html
<html>
<head>
<meta name = "viewport" content = "initial-scale = 1.0,
user-scalable = no">
<link href="style.css" rel="stylesheet" type="text/css">

<script type="text/javascript" src="NKit.js"></script>

<script type="text/javascript">
var navController = new NKNavigationController();
navController.setTitle("Piano solo");
navController.setTintColor(52, 80, 224);
</script>

</head>

<body>
<audio src="song.mp3" controls></audio>
<img src="photo.jpg" width=100%">
</body>

</html>
```

When you test it in Simulator, you'll have something with both visual and musical interest (**Figure 7.2**).

7.2 The revised example with a title bar, photograph, and the audio controls sized and positioned with CSS.

Truly, that's all there is when it comes to adding audio with HTML5!

Incorporating audio with NKAudioPlayer

Another method for incorporating audio into a NimbleKit-based iOS app is by using the **NKAudioPlayer** library item. Doing this shifts the coding work from HTML and CSS to JavaScript and results in a slightly different presentation.

To walk through this example, start with a new Xcode project file and add the song.mp3 and photo.jpg files to the project.

Next, create an instance of **NKAudioPlayer** in the main.html page. Give it the name **player** and have it load the **song.mp3** file:

```
var player = new NKAudioPlayer();
player.loadFile("song.mp3");
```

Leaving this as is does not result in any music playing, however. You also need to create some controls. And strangely, there is no native Objective-C audio controller for NimbleKit to call on, so you need to build one yourself.

For this purpose, I suggest using the NKToolBar library item. It's similar to the **NKTabBarController** that we explored in Chapter 4 except that it has no tabs and **NKToolBar** is 44 pixels instead of 49 pixels tall (or, the same height as the title bar in **NKNavigationController**). Instead of the tabs, **NKToolBar** inserts buttons onto the bar.

The next modification to main.html is to create a toolbar object and name it **audiocontrols**. In addition to naming the instance, it needs to be positioned on the screen on the y-axis. If the screen is going to have a title bar at the top, doing the math tells us that the top of the toolbar should be positioned 372 pixels from the origin (upper-left corner) in order for it to be flush to the bottom of the screen: 480 (full screen height) – 20 (status bar) – 44 (title bar) – 44 (tool bar) = 372. Here's that position of the **NKToolBar** object being set through the first argument passed to the **init()** method:

```
var audiocontrols = new NKToolBar();
audiocontrols.init(372);
```

Next, add the buttons you'll need to restart, play, and pause the audio file. To do this, use the method **toolBar.addButton**. Also note that by default, **NKToolBar** adds buttons starting from the left. Rather than do that, I encourage you to center the buttons so that it looks more like the iPod's audio control. To make this happen, an odd little command is required: **audiocontrols.addFlexibleSpace()**:

```
audiocontrols.addFlexibleSpace();
audiocontrols.addButton("Restart", "", "buttonPressed1");
audiocontrols.addButton("Play", "", "buttonPressed2");
audiocontrols.addButton("Pause", "", "buttonPressed3");
audiocontrols.addFlexibleSpace();
audiocontrols.setStyle("blacktranslucent");
audiocontrols.show();
```

Note that for every **audiocontrols.addButton** method, there are three parameters:

- button label
- button graphic
- callback (JavaScript action)

So, in this example, the three buttons will be labeled Restart, Play, and Pause; there are no graphics assigned to the buttons; and there are three

JavaScript functions named `buttonPressed1`, `buttonPressed2`, and `buttonPressed3`.

Next, let's define the three callback functions:

```
function buttonPressed1()
{
  player.play();
}


function buttonPressed2()
{
  player.resume();
}


function buttonPressed3()
{
  player.pause();
}
```

Note that for this implementation, I didn't actually assign "play" to the play button. What I wanted to do instead is approximate a familiar audio player as closely as possible, so I made the first button the equivalent of restart (rewind to the beginning of the track and play); the second button is labeled "play" but assigned "resume" (so it will play from any point in the track, not just the beginning); and the third is both labeled and assigned "pause."

So here is the final code for main.html:

```
<html>
<head>
<meta name = "viewport" content = "initial-scale = 1.0,
user-scalable = no">
<link href="style.css" rel="stylesheet" type="text/css">
<script type="text/javascript" src="NKit.js"></script>

<script type="text/javascript">
var navController = new NKNavigationController();
navController.setTitle("Piano solo");
navController.setTintColor(52, 80, 224);

var player = new NKAudioPlayer();
```

```
player.loadFile("song.mp3");

var audiocontrols = new NKToolBar();
audiocontrols.init(372);
audiocontrols.addFlexibleSpace();
audiocontrols.addButton("Restart", "", "buttonPressed1");
audiocontrols.addButton("Play", "", "buttonPressed2");
audiocontrols.addButton("Pause", "", "buttonPressed3");
audiocontrols.addFlexibleSpace();
audiocontrols.setStyle("blacktranslucent");
audiocontrols.show();

function buttonPressed1()
{
  player.play();
}

function buttonPressed2()
{
  player.resume();
}

function buttonPressed3()
{
  player.pause();
}

</script>
</head>

<body>
<img src="photo.jpg">
</body>
</html>
```

There is only one CSS rule for the **styles.css** file in this example, to reset the screen's margin and padding to zero:

```
body {
  margin: 0px;
  padding: 0px;
}
```

Figure 7.3 shows the result in Simulator.

7.3 An audio player created with NKAudioPlayer and NKToolBar with labeled buttons.

This isn't bad. But there may be a better way.

The better way that I sought was to graphically look more like the iPod application on an iPhone. So I made restart, play, and pause buttons for the toolbar (**Figure 7.4**).

7.4 Restart, play, and pause buttons.

Now instead of the labels, the JavaScript that pulls in the buttons needs to pull in the PNG images I made: resume.png, play.png, and pause.png. This is how to make that modification:

```
var toolBar = new NKToolBar();
toolBar.init(372);
toolBar.addFlexibleSpace();
toolBar.addButton("", "resume.png", "buttonPressed1");
toolBar.addButton("", "play.png", "buttonPressed2");
toolBar.addButton("", "pause.png", "buttonPressed3");
toolBar.addFlexibleSpace();
toolBar.setStyle("blacktranslucent");
toolBar.show();
```

You could even leave the button labels in the code; the filled image parameters trump the labels, though, so the result would be the same. This is shown in **Figure 7.5**, with the standard labeled buttons replaced with the PNG images.

7.5 The revised NKAudioPlayer and NKToolBar audio screen, now with buttons similar to the iPod application.

So those are some fundamental building blocks for using audio in a NimbleKit-powered iOS application. Be creative, build on these ideas, and have fun with audio!

Summary

This chapter has demonstrated that adding audio to an iOS app is quite straightforward:

- You can use HTML5's audio element, which gives you more flexibility with screen layout.

- Or you can use NimbleKit's NKAudioPlayer library item, which works best when coupled with a toolbar and custom buttons.

Next, you'll learn that working with video is comparable to working with audio: There are both HTML5 and NimbleKit library item options.

8 FOCUS ON APP CONTENT: VIDEO

At the moment, Adobe Flash is still the dominant platform for delivering video on the World Wide Web. But you are likely familiar with the famous Steve Jobs public letter about Flash from spring 2010 that drew a clear line in the sand between iOS devices and Flash.[1] So until Flash is supported in iOS, you need to use other alternatives for playing video.

Fortunately, the approach to video in iOS apps is similar to the approach to audio: There are HTML5 and NimbleKit library item options. Let's take a look and see how video works in each case.

Delivering video with HTML5 on iPad

Playing a video file with HTML5 is about as easy as playing an audio file, but here's the caveat: At least with the NimbleKit Objective-C framework, the HTML5 option is only fully functional on the iPad and not on Apple's pocket-sized iPhone or iPod touch.

The limitation on iPhone and iPod touch is this: The embedded HTML5 video player can be displayed in a screen view, but when you touch the play control a native iOS video player interface is displayed. Essentially, it's the same phenomenon that happens in the YouTube app that comes with these devices.

On the iPad, the HTML5 video player actually plays the video embedded, right where it is. To use the HTML5 option, just implement its new `video` element and have an MPEG-4 (.m4v) video file available. Here's the format for the HTML markup:

```
<video src="name.m4v" width="x" height="y"
poster="name.jpg" controls></video>
```

Let's explore these options in detail:

- The `src` attribute identifies the video file to be played (and don't forget to **Add to Project** to copy the file to your Xcode project's HTML directory, otherwise your app will not locate the video file!).

- As with the `object` tag it's replacing, it's a good habit to continue specifying the `width` and `height` attributes of your video. This keeps the black play area, or stage, the same size and aspect ratio as your video.

- The `poster` attribute allows you to specify an image file to use as the poster frame or placeholder for the stage prior to playing. Either a JPG or PNG can be used.

- Finally, you need to add the `controls` attribute to display standard controls. Without this attribute, the user can't play the video.

Designing a sample video application

Armed with this knowledge, let's design a sample app to demonstrate the `video` tag in action on the iPad. As it happens, I have a short video clip of one of my daughters practicing the piano that I can share. (Download it, along with the other related items, at **iosapps.tumblr.com**.) I'll make this example an app that teaches people about the piano. The introductory screen will feature a title bar, the sample video clip, and the first three paragraphs of text content about the piano from Wikipedia:[2]

```
<p>The piano is a musical instrument played by means of a
keyboard. It is widely known as one of the most popular
instruments in the world. Widely used in classical music
for solo performances, ensemble use, chamber music, and
accompaniment, the piano is also very popular as an aid to
composing and rehearsal. Although not portable and often
expensive, the piano's versatility and ubiquity have made
it one of the world's most familiar musical instruments.
</p>

<p>Pressing a key on the piano's keyboard causes a felt-
covered... ...classification, pianos are grouped with
chordophones.</p>

<p>The word piano is a shortened form of the word
pianoforte...
...speed with which the hammers hit the strings.</p>
```

And I have not forgotten that the point of this exercise is to work with video, so **Figure 8.1** shows a sample clip, **clip.m4v**, as viewed in QuickTime Player.

8.1 Clip.m4v in QuickTime Player on the Mac desktop.

Let's start this exercise by adding the text content and video to the main. html file in a new NimbleKit-based Xcode project for the iPad. As with the previous exercise, remember to add the video file asset to the project's HTML folder first.

The piano content looks like this as HTML, with the **NKNavigationController** added to the head to create a title bar and the **video** tag added to pull in the video.

NOTE **Landscape orientation**

You are adding an additional JavaScript function here that references NimbleKit's **NKIsPageSupportsAutoOrientation** to enable landscape orientation, which is critical for an iPad app. According to official iPad specs, an iPad app will *not* be approved unless it supports both portrait and landscape orientations.

```html
<html>

<head>
<meta name = "viewport" content = "initial-scale = 1.0,
user-scalable = no" />

<script type="text/javascript" src="NKit.js"></script>

<script type="text/javascript">

var navController = new NKNavigationController();
navController.setTitle("iPad Video Example");
navController.setTintColor(0, 102, 255);

function NKIsPageSupportsAutoOrientation()
{
return "yes";
}

</script>

</head>

<body>

<h1>The Piano</h1>
```

```
<video src="clip.m4v" width="320" height="240" controls>
</video>

<p>The piano is a musical instrument played by means of a
keyboard. It is widely known as one of the most popular
instruments in the world. Widely used in classical music
for solo performances, ensemble use, chamber music, and
accompaniment, the piano is also very popular as an aid to
composing and rehearsal. Although not portable and often
expensive, the piano's versatility and ubiquity have made
it one of the world's most familiar musical instruments.
</p>

...two more paragraphs of text here...

<p><em>Source &mdash; http://en.wikipedia.org/wiki/Piano
</em></p>

</body>

</html>
```

Testing this in Simulator looks like the screen shown in **Figure 8.2**.

8.2 A "rough draft" of an iPad app screen with embedded video.

There are still several opportunities to improve the presentation of this embedded video. First, let's add a poster image to replace the black background behind the video control overlay. **Figure 8.3** shows the photo. jpg file to be added to the project's HTML folder.

8.3 The photo.jpg (320 × 240 pixels).

Highlighted in the following code is a modification whereby the `video` element finds the poster image and displays it:

```
<video src="clip.m4v" width="320" height="240"
poster="photo.jpg" controls></video>
```

Note also that iOS automatically overlays a white Play button on top of the poster image (**Figure 8.4**). Thanks, iOS!

I'm sure you'd like to style this content a bit more (I know I would)—and what about making the presentation more portrait and landscape friendly? Fortunately, both can be done with CSS! So let's create a new file, video.css, and link to it from main.html:

```
<link href="video.css" rel="stylesheet" type="text/css">
```

Now you can style the video element. My suggestion is to try these properties as noted in the following code:

```
video {
  display: block;
  margin-right: auto;
  margin-left: auto;
  margin-top: 0px;
  margin-bottom: 30px;
  -webkit-box-shadow: 10px 10px 10px #888;
  box-shadow: 10px 10px 10px #888;
  padding: 15px;
  border: 1px solid #ccc;
}
```

What is happening here? By displaying as block and having auto left and right margins, the video is centered horizontally—even when the screen is rotated. So file this technique under "iPad design tips" and notice that it borrows from techniques used to center things in flexible-width and liquid layout web pages.

The other notable styling element here is the CSS3 `box-shadow` property (with proprietary `-webkit-` prefix). Combined with the border, margin, and padding properties, the result is a nice frame for the video. It's HTML5 and CSS3 working together—and you haven't even gotten to the chapter that focuses on these magical specifications! (Consider this just a preview.)

NOTE Prefixed CSS properties

With prefixed CSS properties it's good practice to follow it with the non-prefixed property for forward-compatibility. Someday, Apple may stop supporting `-webkit-box-shadow` and switch over to just `box-shadow`. It feels redundant to do so, but could save some pain later as CSS matures and browser prefixes are less necessary.

Next, use the following rules to add a margin to the entire screen and style the h1—yes, I'm actually using Zapfino:

```
body {
  background-color: #fff;
  margin: 20px;
  padding: 0px;
  font-family: Helvetica;
}

h1 {
  margin-top: 50px;
  margin-bottom: 10px;
  padding: 0px;
  font-size: 3em;
  font-family: Zapfino;
  line-height: 2em;
  text-align: center;
}
```

With the new styling in place, the results are much more presentable and attractive (**Figure 8.4**).

8.4 Our styled HTML5 video app on the iPad.

And the layout adjusts to keep everything centered when rotated to landscape (**Figure 8.5**)!

8.5 A landscape view of the HTML5 video app on the iPad.

So that's how it's done on the iPad with HTML5's **video** element. And remember that it also works on the iPhone and iPod touch. Well, sort of. On those devices, it triggers a native iOS player, so it doesn't play the video embedded in the screen.

```
{
return "yes";
}
```

```
</script>
```

The `NKNavigationController` library item and `NKIsPageSupports-AutoOrientation` function should look familiar from previous exercises, but I've highlighted the portion that is new. These two lines create a new instance of `NKVideoPlayer` called `videoPlayer`, and instruct it to open the same video file that we used in the previous exercise, clip.m4v.

Now you need to design a mechanism for telling `NKVideoPlayer` to play clip.m4v and call up the iOS video player overlay mechanism. The technique you will use is a CSS-styled button similar to the example in Chapter 6. Here is some HTML that can be added to the body of the main.html file:

```
<ul class="button">
<li><a href="#" onclick="videoPlayer.play()">Play video</a></li>
</ul>
```

And to style the list item to appear as a native-looking button with rounded corners and text shadow, here is the CSS:

```
.button {
  font-size: 14pt;
  display: inline;
}

ul.button {
  list-style-type: none;
}

ul.button li a {
  color: #fff;
  background-color: #06f;
  width: 220px;
  text-align: center;
  padding: 8px 0px 0px 0px;
  height: 32px;
  border: 1px solid #666;
  display: block;
  -webkit-border-radius: 10px;
```

```
  border-radius: 10px;
  text-shadow: 0px 1px 0px #ccc;
  text-decoration: none;
  margin-top: 20px;
  margin-right: auto;
  margin-left: auto;
}
```

To maintain the spirit of our previous example, you can use the same title and image as in the poster image for the HTML5 player; just insert this before the button code:

```
<h1>The Piano</h1>
<img src="photo.png" width="100%" />
```

And let's maintain the same basic look, too, by styling the **h1** with Zapfino again. (I mean, how often can you use Zapfino and get away with it? We may be pushing the limit with this book's examples!) Here are the CSS properties to add to the video.css file for styling the **h1** element:

```
h1 {
  margin-top: 20px;
  margin-bottom: 10px;
  padding: 0px;
  font-size: 2em;
  font-family: Zapfino;
  line-height: 2em;
  text-align: center;
}
```

Here is the complete main.html file, with linked stylesheet:

```
<html>
<head>
<meta name = "viewport" content = "initial-scale = 1.0,
user-scalable = no" />

<link href="video.css" rel="stylesheet" type="text/css" />

<script type="text/javascript" src="NKit.js"></script>

<script type="text/javascript">
var navController = new NKNavigationController();
navController.setTitle("iPhone Video Example");
navController.setTintColor(0, 102, 255);
```

```
var videoPlayer = new NKVideoPlayer();
videoPlayer.openFileName("clip.m4v");

function NKIsPageSupportsAutoOrientation()
{
  return "yes";
}

</script>

</head>
<body>
<h1>The Piano</h1>
<img src="photo.png" width="100%" />
<ul class="button">
<li><a href="#" onclick="videoPlayer.play()">Play video</
a></li>
</ul>

</body>
</html>
```

Testing this in Simulator, the app will look like the example in **Figure 8.7**.

8.7 The NKVideoPlayer app example on the iPhone.

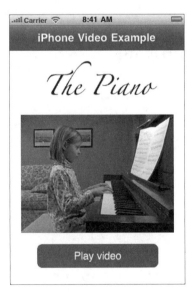

When you click the Play video button, the video overlay appears with a
native iOS controller. (**Figure 8.8**)

8.8 The NKVideoPlayer in action,
showing the native video player overlay.

Do you recall that I said this universal app example will run full screen on
the iPad? If you've been testing in Simulator using the iPhone setting, take
the app for a spin with it set to iPad. Just go to Xcode's Overview menu and
change Active Executable to iPad Simulator as shown in **Figure 8.9**.

8.9 Testing a universal app on
the iPad requires telling Xcode to
do so via the Overview menu's
Active Executable setting.

Now when you Build and Run, the version that you get on the iPad (**Figure 8.10**) no longer plays in the small iPhone window like iPhone apps do. It sizes up (though does not scale) to the iPad's screen size of 768 × 1024 pixels.

8.10 The NKVideoPlayer app example on the iPad.

And note that when you click the Play video button, the same video player comes up and the video plays full screen, both in portrait orientation. (**Figure 8.11**) and landscape orientation (**Figure 8.12**).

8.11 The NKVideoPlayer in action on the iPad, showing the native video player overlay and video sized to full screen (width).

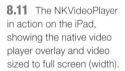

8.12 The same sample app running on the iPad after the device orientation is rotated to landscape.

And that is how you design an iOS app screen using `NKVideoPlayer`!

Summary

In this chapter you learned how to

- Embed MPEG-4 (.m4v) video files into iPad app screens using the new HTML5 `video` tag (while learning that the videos will work, but behave differently, on iPhones and iPod touches).

- Use NimbleKit's `NKVideoPlayer` library item to provide a video player experience that is just like the YouTube app, with a video player overlay instead of embedded video.

- Make a universal app that plays full screen on iPhones, iPod touches, and iPads—from the same app binary, and thus also from the same iTunes App Store download/purchase.

- Style universal and iPad app layouts to accommodate landscape orientation.

Now that I've introduced you to some methods of incorporating text, image, audio, and video content into iOS apps, the next chapter focuses more closely on some additional ways of using HTML5 and CSS3 in your iOS app designs.

REFERENCES

1. http://www.apple.com/hotnews/thoughts-on-flash/
2. http://en.wikipedia.org/wiki/Piano

Delivering video with NKVideoPlayer

This next method of incorporating video uses NimbleKit's `NKVideoPlayer` library item. It differs from the HTML5 video element in that it doesn't have an embedded play mode. Instead, playing a video with `NKVideoPlayer` calls up a native video player overlay, and this behavior is consistent across all three iOS devices.

To see how this works, use the same assets from the previous example but change it up a bit at the point where you create a new Xcode project. This time, select iPhone/iPad as shown in **Figure 8.6** to create what is called a *universal app*—that is, one app that will run natively in full-screen mode on the iPad as well as on the iPhone and iPod touch.

8.6 Creating a universal app that will run full screen on iPad, iPhone, and iPod touch.

Aside from making this choice when you start the new project, there's nothing else to do differently until you test the app in Simulator.

After Xcode opens the new project, you can begin editing main.html right away by adding the following JavaScript to the head of the page:

```
<script type="text/javascript">
var navController = new NKNavigationController();
navController.setTitle("iPhone Video Example");
navController.setTintColor(0, 102, 255);

var videoPlayer = new NKVideoPlayer();
videoPlayer.openFileName("clip.m4v");

function NKIsPageSupportsAutoOrientation()
```

9 HTML5 AND CSS3

The buzz surrounding HTML5 and CSS3 has definitely been getting louder in the past few years. But for web designers who still need their sites to work consistently and predictably across all major browsers—antique as well as modern—questions remain: What is safe to use now, and what is still premature and unevenly supported? It can be a bit frustrating—we want to start learning how it all works and gain some experience with HTML5 and CSS3, but not push the envelope too far, too fast.

Fortunately, the browser engine in Apple's mobile operating system uses WebKit, and the same great support for most new HTML5 and CSS3 rules in mobile Safari is also available for you to leverage in native iOS apps built with the NimbleKit Objective-C framework and its web view.

Let's take a closer look at how designing iOS apps gives you the freedom to use HTML5 and CSS3—not tomorrow but today, and with no risk at all.

Exploring additional HTML5 elements

Chapter 5 introduced you to formatting content in NimbleKit apps with well-established HTML markup such as heading and paragraph tags, unordered lists, and definition lists. But what about the new HTML5 elements that can enhance the semantic meaning of our markup? Are iOS apps a safe place to start becoming familiar with these?

Yes!

Let's begin with some easy-to-adopt structural content elements.

Section

Adopting the `section` element is a fantastic way to safely wade into the new HTML5 waters, as it provides an immediate win: It helps to extricate you from having markup that has way too many `div` tags.

Think about sections in any printed content (this book, for instance), and you can intuitively figure out how this new element can help you with web and app content. In the case of print, sections of a book are indicated with a different typographical treatment because the paragraphs, figures, and notes in between these section indicators are related groups of information. Change topic or area of focus, and insert a new indicator to communicate that you've entered a new section.

Using `section` in your HTML is just as intuitive. In fact, I would argue that using `section` makes you a better designer because it forces you to become familiar with the meaning of the content you are marking up.

Let's use the U.S. Constitution as an example, since we talked earlier about the U.S. Bill of Rights.[1] Using HTML4, you could begin marking up Article I of the Constitution like this:

```
<div>

<h1>Article I - The Legislative Branch</h1>

<h2>Section 1 - The Legislature</h2>
<p>All legislative Powers herein granted shall be vested
in a Congress of the United States, which shall consist of
a Senate and House of Representatives.</p>

<h2>Section 2 - The House</h2>
```

```
<p>The House of Representatives shall be composed of
Members chosen every second Year by the People of the
several States, and the Electors in each State shall have
the Qualifications requisite for Electors of the most
numerous Branch of the State Legislature...</p>

<h3>Authors</h3>
<ul>
<li>James Madison</li>
<li>George Washington</li>
...
</ul>

</div>
```

And there is certainly nothing wrong with that—it is still valid as HTML5. The only problem with the `div` is that by itself, it has no semantic meaning. As the W3C HTML specification notes, it is "a generic mechanism for adding structure to documents" and relies on a `class` or `id` to help it out with presentation.[2] So it's really the HTML equivalent of a brown paper wrapper. Of course, you could `class` the `div`:

```
<div class="article">
```

But even here, classing the `div` as an article is technically subjective: It does not add semantic meaning because it is serving only as a stylesheet selector. However, `section` does two things: It groups the content like `div`, and communicates that the content inside belongs together in a logical way:

```
<section>

<h1>Article I - The Legislative Branch</h1>

<h2>Section 1 - The Legislature</h2>
<p>All legislative Powers herein granted shall be vested
in a Congress of the United States, which shall consist of
a Senate and House of Representatives.</p>

...

</section>
```

And naturally, you can still `class` or `id` a section for presentational purposes.

Header and footer

Like `section`, `header` and `footer` can help alleviate the too-many-divs problem that HTML4 pages of content can incur while adding more semantic clarity to your markup. I'm covering them here together because they share some properties, both intuitively and nonintuitively.

Intuition would lead us to believe that content tagged with `header` would be high-level, introductory, and perhaps also navigational content at the top of a page; similarly, `footer` would be at the bottom of a page and contain reference information about the content above it: copyright, date of authorship, author, and perhaps contact information.

Fortunately, our intuition is correct… mostly. The only thing that might be tricky to get used to is that you're not limited to one `header` or `footer` per page, and they are not required to be at the top and bottom of a page respectively (though one suspects that this is where most designers will use them). So to modify our Constitution markup:

```
<section>

<header>
<h1>Article I - The Legislative Branch</h1>
</header>

<h2>Section 1 - The Legislature</h2>
<p>All legislative Powers herein granted shall be vested
in a Congress of the United States, which shall consist of
a Senate and House of Representatives.</p>

...
<footer>
<h3>Authors</h3>
<ul>
<li>James Madison</li>
<li>George Washington</li>
...
</ul>
</footer>
</section>
```

See how the more HTML5 you use, the more semantic structure your content takes on? Also note that a `section` may contain a `header` or `footer`.

Article

Where things get slightly hazy is with the `article` element. It is intended to wrap areas of "self-contained content". Well, what does that mean, exactly, and how does that differ from a `section`?

The metaphor I picked up from John Allsopp, author of *Developing with Web Standards* (New Riders, 2009), that I find extremely helpful for deciphering this apparent conundrum is this: `article` wraps content that could existing meaningfully and independently from the context of the whole, whereas `section` is literally a section of something—it only makes sense within its context.[3]

The comparison that Allsopp makes is

- A newspaper article typically makes complete sense on its own, so it may be wrapped with the `article` element.

- A traditional book chapter (that is, one that is not a short story or an article in a compilation) relies on the chapters before and after it to make sense—it's part of a series of items. This would be wrapped in a `section` element.

Think of `article` as a more specialized `section` when you're considering its use. If a suspected `section` of content can stand on its own, consider using `article` instead.

Aside

Consider the `aside` to be another specialized form of `section` and, as with `article`, trust that its name is laden with intentional meaning. An aside could be

- A tangential reference, such as the tips and notes in this book

- A pull quote that you want to set off semantically, and likely wish to style uniquely

One way to measure whether content fits the intentions of `aside` is to see whether the context in which it resides makes sense without it. If it does, the content in question is likely tangential enough to be an `aside`.

Nav

I saved **nav** for the end of the structural element discussion because it may be the least likely to be used in an iOS app setting. The reason for this is that a lot of app navigation is in the JavaScript portions of the code, so that wouldn't be wrapped by **nav** tags.

On the other hand, any buttons or unordered list navigation elements in an HTML page could certainly be wrapped in **nav**.

Next I'll give a quick nod to a few more HTML5 elements, two that we've covered already and one that will be mentioned briefly. These fall under the category of rich media.

Audio and video

The **audio** and **video** HTML5 elements were introduced in Chapters 7 and 8, and they work with Xcode and NimbleKit on iOS devices. Just remember that NimbleKit offers alternative classes in its library, NKAudioPlayer and NKVideoPlayer, which give you slightly different user experiences and options.

Canvas

The **canvas** element allows designers and developers to draw lines and shapes, fill shapes and areas, and animate objects. It's a powerful, JavaScript-driven element that happens to be well supported by Xcode and NimbleKit.

Two awesome examples on the web that could get you very excited about **canvas** are the Visualizing the World Cup[4] and Visualizing the Stanley Cup[5] websites. Pay them a visit in either Safari on a Mac or mobile Safari on an iOS device. All of the visualization and typography (aside from the team logos and background images) use HTML5 and CSS3.

Very impressive!

However, the **canvas** element is where I need to draw the line (no pun intended) in the content I attempt to cover with any respectable depth. I didn't want to omit **canvas** completely and am pretty impressed with what it can do. On the other hand, **canvas** doesn't really do anything on its own—it merely opens the door for a ton of additional uses of JavaScript.

The main reason `canvas` isn't demonstrated here is because it's a fairly complicated way to do some otherwise simple things. This method works well when it scales to a large enough project, and when dynamic image generation makes the most sense. An example would be app content that's driven by a large database, where the visualization is more efficiently done on the fly so that servers don't need to be burdened with a bunch of prefabricated image files.

Such a scenario certainly makes `canvas` useful—except that this book doesn't cover large application development. And I cannot recommend that you use `canvas` to draw simple shapes, either; that's a case where the HTML5 route is actually longer and more complicated than creating bitmapped image files.

Additional HTML5 elements

The HTML5 elements covered in this book should be quite useful to you as web designers, and the most relevant to iOS app design with NimbleKit. But they are by no means all of the HTML5 elements. If you want to read up on all of them, visit the clever Periodic Table of the (HTML5) Elements, brought to you by Josh Duck[6] (**Figure 9.1**).

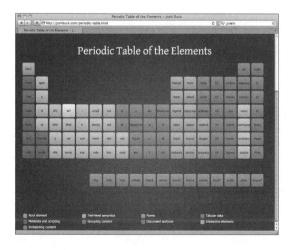

9.1 The Periodic Table of the (HTML5) Elements. Brilliant!

More design options with CSS3

CSS3 allows us to quickly take a pedestrian design and elevate it to something special, and do it in a way that is extremely agile and malleable. In other words, the process of using CSS3 for presentational and visual effects enables adjusting and tweaking on-the-go as well as revisions and even substantial redesign much later. Let's explore a few examples to see how CSS3 helps us in this way—and realize that there are many more CSS3 techniques to explore by checking out the actual specification or reading books that focus specifically on the topic.

Border-radius

The `border-radius` property was introduced in Chapter 5 (see Working with images and Figure 5.9):

```
img {border-radius: 15px;}
```

and in Chapter 6 (see Method two: Styling an HTML button and Figure 6.2):

```
ul.button li a {border-radius: 10px;}
```

In addition to being applied to HTML elements, it can be applied to an entire div class or id:

```
#box {border-radius: 10px;}
```

And to get a bit crazier with it, you can specify different radii for different corners:

```
#crazybox {
  border-bottom-left-radius: 10px;
  border-bottom-right-radius: 8px;
  border-top-left-radius: 12px;
  border-bottom-right-radius: 2px;
}
```

But it gets crazier still: You can supply two values, the first for the horizontal distance and the second for the vertical distance, allowing you to create something that is more elliptical:

```
#ellipse {
  border-radius: 50px 25px;
}
```

That gives you a bit more to explore with `border-radius`. Have fun!

RGBA

CSS has always supported color very well. But prior to CSS3 and Red-Green-Blue-Alpha (RGBA), transparency was a hack that was accomplished by using image files that were saved with transparency. This method works OK, but it requires a lot of time-consuming image production. Worse yet, to make any changes to a design that uses a lot of images with transparency, you have to rework all of the images and resave them as PNGs.

Fortunately, RGBA is a much more agile and efficient way to achieve the same results, allowing you to tweak the design or change out images quickly without having to redo a bunch of image editing. So how does this work, and how does it result in something useful for iOS apps?

RGBA is an extension of the RGB color model that is used in many places—television and computer screens, the Photoshop Color Picker, and the NKNavigationController in NimbleKit to name a few. The first three values (R, G, and B) can be between 0 and 255 (0,0,0 is black; 255, 255, 255 is white; all other hues are made using combinations that lie in between these values). The A value can only be between 0 and 1, with 1 being completely opaque, 0 being completely transparent, and decimal values in between doing the real magic.[7]

Using RGBA is probably most practical when applied to a background. Here's a background setting of 100 percent black with 50 percent opacity:

```
background: rgba(0,0,0,.5);
```

So how might one apply this effect when designing an iOS app? Let's use a farmers' market app as an example. Bear with me—the screen does a good job of making vegetables look sexy.

Anyway, imagine designing an app for a local farmers' market. You might have several of these screens, each one focusing on different items: vegetables, jams and jellies, local meats, and so on. This example will be a vegetables screen with three items: beets, carrots, and kale.

Start by making a new Xcode project for iPhone and modifying the main.html file to match the code in Figure 9.2. What you're doing is adding an unordered list containing the three vegetables. As with the other examples, you can download this in its entirety from iosapps.tumblr.com.

(This is particularly useful in this case because then you can use the same image files that I created.)

```html
<html>
<head>
<meta name = "viewport" content = "initial-scale = 1.0,
user-scalable = no">
<link href="style.css" media="screen" rel="stylesheet"
type="text/css">

<script type="text/javascript" src="NKit.js"></script>

</head>
<body>

<ul class="food">
<li>
<img src="beets.jpg">
<strong>Beets</strong>
<em>$4.49</em>
</li>
<li>
<img src="carrots.jpg">
<strong>Carrots</strong>
<em>$3.99</em>
</div>
</li>
<li>
<img src="kale.jpg">
<strong>Kale</strong>
<em>$2.99</em>
</li>
</ul>

</body>
</html>
```

Now that we have some HTML underway, let's get some CSS started on another burner. Create a styles.css file and add it to the project with the following rules:

```css
body {
  margin: 0px;
}
```

```
ul.food {
  width: 320px;
  margin: 0px;
  padding: 0px;
  list-style-type: none;
}
```

These rules alone won't do much for this app—yet. The point is to establish a baseline for the design. With the above HTML and CSS files in the Xcode HTML folder, testing the app in Simulator should give you the result shown in **Figure 9.2**.

9.2 Nice photos, but this app's UI could use a bit more work.

Now the fun begins. As I was reviewing some of the books that have taught me some solid skills, I was reminded of a great RGBA example in *Hand-crafted CSS* by Dan Cederholm. My apologies to Dan for my implementation—his designs are always so elegant, so we'll see whether my adaptation of his method for the iOS screen can approach his level of CSS3 craft. (It may not, but at least now you know that I've chosen the right goal to work toward!)

With Dan setting the bar, let's give the unordered list with the .**food** class some more definition by defining its **li** tags like so:

```
ul.food li {
  position: relative;
  width: 320px;
  height: 140px;
  margin: 0px;
  padding: 0px;
}
```

This sets up the `li` to also be page-width and the `position: relative` sets you up to create a span for wrapping the content like so:

```
<li>
<div class="veggies">
<img src="beets.jpg">
<span>
<strong>Beets</strong>
<em>$4.49</em>
</span>
</div>
</li>
```

This enables you to style the name and price content as an overlay that will be positioned right on top of the image, flush bottom. Here is the CSS for that effect:

```
div.veggies span {
  display: block;
  position: absolute;
  width: 100%;
  height: 30px;
  bottom: 0;
  left: 0;
  padding: 5px;
  font-family: American Typewriter;
  font-size: 1.5em;
  color: #ccc;
  background: rgba(0,0,0,.5);
}
```

You're doing several things here: absolute positioning the block above the image, setting the width of the overlay to 100 percent, making its height 30 pixels, and giving the text 5 pixels of padding. And as long as you're giving the label some style, you're getting away from the default typeface by setting the font to American Typewriter, sizing it up to 1.5 ems, and making the text white.

This will all look very nice on top of the most important section (high-lighted): the black background with 50 percent opacity.

Now create a title bar at the top of the page by adding the following to the head of main.html, and use your newfound RGBA knowledge to pay particular attention to what's in the NKNavigationController JavaScript:

```
<script type="text/javascript">
var navController = new NKNavigationController();
navController.setTitle("Vegetables");
navController.setTintColor(0, 0, 0);
</script>
```

I know, I cheated and highlighted it. But see the 0, 0 ,0 ? Hopefully you remember this from Chapter 4 (see Implementing the title bar) — the RGB values that NimbleKit uses to define the background color of the title bar work the same way they do in our CSS!

After amping up the HTML and CSS with the above code, and testing anew in Simulator, you should get the results shown in **Figure 9.3**.

9.3 The restyled vegetables screen, now featuring a navigation controller at the top of the page and image labels that have a nice translucent effect thanks to RGBA.

It's great to know that if you have experience styling with some of the newer CSS3 specifications like RGBA (and border-radius, already used in Chapter 6), you can use these skills in NimbleKit-powered iOS applications.

@font-face

Another great CSS3 tool to add to your iOS toolbox is @font-face. The @font-face rule is an invigorating breath of fresh typographic air for web designers and, fortunately, what works for modern web browsers also works (in a more limited fashion) for designing iOS apps with HTML, CSS, and the NimbleKit Objective-C framework.

Technically @font-face dates to the CSS2 specification, but it didn't really take hold until CSS3. As the W3C specification notes, @font-face allows designers to go beyond the traditional set of "web-safe" fonts, and to do so in a way that is consistent across any modern browser (or browser engine) that supports CSS3.[8] This gives designers a level of independence with rendered HTML typography that we have dreamed about for years!

But before you launch into using @font-face, you should know what fonts are already resident in the iOS (**Figure 9.4**).

9.4 The fonts that come installed in iOS on the iPhone, iPod touch, and iPad.

American Typewriter
  abcefgijop 123 ABC

American Typewriter Condensed
  abcefgijop 123 ABC

Arial
  abcefgijop 123 ABC *abcefgijop*

Arial Rounded MT
  abcefgijop 123 ABC

Courier New
  abcefgijop 123 ABC *abcefgijop*

Georgia
  abcefgijop 123 ABC *abcefgijop*

Helvetica
  abcefgijop 123 ABC *abcefgijop*

Marker Felt
  abcefgijop 123 ABC *abcefgijop*

Times New Roman
  abcefgijop 123 ABC *abcefgijop*

Trebuchet MS
  abcefgijop 123 ABC *abcefgijop*

Verdana
  abcefgijop 123 ABC *abcefgijop*

Zapfino
  abcefgijop 123 ABC

Not a terribly exciting selection, is it?

It's easy to see why an iOS app designer would be interested in using another typeface that isn't represented in this list. So how do we go about expanding our options beyond these meager staples? The basic formula for the rule is

```
@font-face { <font-description> }
```

and the description has the same form as other CSS rules, with the familiar syntax

```
descriptor: value;
```

Now all we need is a new typeface. Hmm—where do we find fonts to install so we have more typographic freedom with our HTML content?

Kernest is one such place.

Kernest

Kernest is a web font services site that delivers browser-supported font files in several formats including Scalable Vector Graphic (SVG) format, the format that's supported by the web rendering engine in iOS. Kernest was launched in July 2009 by Garrick Van Buren and currently provides 100 fonts for online or local use (**Figure 9.5**)[9].

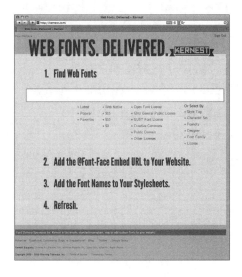

9.5 Kernest, a web font service.

Here are the steps for using a font from Kernest in an iOS app:

1. Search for a new font: Kernest gives you several ways to search through its library of fonts including by style, foundry, designer, font family, or license.

2. Register for the service (this step is free) so you can sign in. Signing in is required to download new fonts.

3. Once you find the font you wish to use, download it and save it to a local drive.

4. Unzip the file archive and add the .svg and .css files that are provided to an open Xcode project.

5. Edit the .css file to reference only the .svg file, and delete the `fonts/` subdirectory from the local url.

6. Add the appropriate class for your new font, as defined in the .css file, to your HTML and you are in business!

Let's walk through an example to see how this works in practice. For this exercise, I visited Kernest and chose an attractive typeface called Juvelo.[10] I found it by searching in the public domain category, because in this case I didn't want to pay to license a font (**Figure 9.6**).

9.6 The Juvelo preview screen in Kernest.

When I download the font files, they arrive in a package called juvelo.zip, which when unzipped looks like **Figure 9.7**.

9.7 The Juvelo font files downloaded from Kernest.

Among these files are two that are needed: juvelo.css and juvelo.svg. Add these to an Xcode project so they appear in the HTML directory under Groups & Files (Project > Add to Project) (**Figure 9.8**).

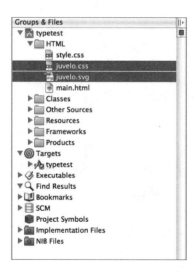

9.8 The Juvelo .css and .svg files added to an Xcode project.

Now we can do some coding. This is where it gets fun!

First, edit the main.html file so you have the following:

```
<html>
<head>
<meta name = "viewport" content = "initial-scale = 1.0,
user-scalable = no">

<script type="text/javascript" src="NKit.js"></script>
```

```
<script type="text/javascript">

var navController = new NKNavigationController();
navController.setTitle("Web fonts FTW!");

</script>

</head>

<body>
<h1>Juvelo</h1>
<hr>
<p>Contributor: Barry Schwartz</p>
<p>Released in 2006</p>
<p>Commercial use permitted</p>
<p>For more information: http://crudfactory.com/font/show/
juvelo</p>
</body>

</html>
```

When you test it in Simulator, you'll get this result—and it's lying because it's not yet typeset in Juvelo but rather Safari's default, Times New Roman (**Figure 9.9**).

9.9 Our text rendered in Times New Roman—not quite for the win yet.

Next, let's edit juvelo.css. The version we downloaded from Kernest looks like this—but I've highlighted what should be deleted from the file so it will work in Simulator and on iOS devices:

```
/* KERNEST.COM WEB FONT CSS GENERATED FOR juvelo */

/*
Juvelo
http://home.comcast.net/~crudfactory/cf3/juvelo.xhtml
Foundry: The Crud Factory, http://home.comcast.
net/~crudfactory/cf3/index.xhtml
Contributors:
License: Public Domain, http://creativecommons.org/
licenses/publicdomain/
*/

@font-face {
  font-family: 'Juvelo';
  src: url('fonts/juvelo.eot');
  src: local(':://'), url('fonts/juvelo.svg#juvelo')
format('svg'), url('fonts/juvelo.woff') format('woff'),
url('fonts/juvelo.otf') format('opentype');
}
.juvelo {
  font-family: 'Juvelo';
  line-height: 140%;
  text-rendering: optimizeLegibility;
}
```

After editing, the results are:

```
/* KERNEST.COM WEB FONT CSS GENERATED FOR juvelo */

/*
Juvelo
http://home.comcast.net/~crudfactory/cf3/juvelo.xhtml
Foundry: The Crud Factory, http://home.comcast.
net/~crudfactory/cf3/index.xhtml
Contributors:
License: Public Domain, http://creativecommons.org/
licenses/publicdomain/
*/

@font-face {
  font-family: 'Juvelo';
```

```
  src: url('juvelo.svg#juvelo') format('svg');
}
.juvelo {
  font-family: 'Juvelo';
  line-height: 140%;
  text-rendering: optimizeLegibility;
}
```

In addition to adding Juvelo to the page, I would like my <h1> to be larger, but I want to keep the non-typeface styling in a separate stylesheet. Do this by creating a style.css file, adding it to the Xcode project, and setting these rules to fine-tune the page:

```
body {
  font-size: .9em;
  margin: 20px;
  margin-top: 40px;
}


h1 {
  font-size: 4em;
}
```

The final step is to edit main.html so it links to both stylesheets, and has a body class of Juvelo so the text is rendered with that typeface. Your HTML should look like this, with these changes highlighted:

```
<html>
<head>
<meta name = "viewport" content = "initial-scale = 1.0,
user-scalable = no">
<link href="juvelo.css" media="screen" rel="stylesheet"
type="text/css">
<link href="style.css" media="screen" rel="stylesheet"
type="text/css">

<script type="text/javascript" src="NKit.js"></script>
<script type="text/javascript">

var navController = new NKNavigationController();
navController.setTitle("Web fonts FTW!");

</script>

</head>
```

```
<body class="Juvelo">

<h1>Juvelo</h1>
<hr>
<p>Contributor: Barry Schwartz</p>
<p>Released in 2006</p>
<p>Commercial use permitted</p>
<p>For more information: http://crudfactory.com/font/show/
juvelo</p>
</body>

</html>
```

Testing the new result in Simulator should give you the following result
(**Figure 9.10**)!

9.10 Our text being rendered in Juvelo—for the win!

Congratulations—you are now free to roam the World Wide Web for new
and interesting SVG format typefaces! Doesn't it feel great?

Gradients

There are also CSS3 effects that aren't technically part of the W3C specification (yet—but they might be eventually). The tactic being used by
some of the browser developers such as Apple and Mozilla is this: If they

build on the existing CSS3 framework by adding new rules and effects, and designers adopt and implement them widely, the W3C will take notice and eventually incorporate the new rules and effects into the specification.

An example of this is CSS gradients, so let's take a look at how they can help enhance our iOS app designs as well as save us some work.

A widely used technique to apply a gradient to a background is to design a narrow sliver of a graphic with a chosen gradient, and specify this image as a background via CSS. The graphic can be just 1 pixel wide, but by repeating it on the x-axis, we can fill the background of a div or another page element. The result is a nice gradient and the performance is pretty high because the load time for such a small graphic is negligible.

However, the work to redesign a gradient created this way is a pain. Just like the earlier example of transparency, methods that tether us to graphics require us to go through laborious updates of graphics when redesigns are called for. Plus, designing the effect the first time is kind of a pain.

Fortunately, CSS gradients are swooping in to save us both time and headache, because it's a lot easier to specify a gradient in a stylesheet. How does it work?

Let's continue working with the same sample Xcode project we just used to explore @font-face and add a background gradient to the page. Let's have the gradient go from white at the top of the page to dark gray at the bottom and go from edge to edge. To do this, just add this to the style.css stylesheet:

```
.linear {
  width: 100%;
  height: 100%;
  background: -webkit-gradient(
    linear,
    left top,
    left bottom,
    from(#fff),
    to(#333));
}
```

Because **gradient** is not formally part of the CSS3 spec yet, but is embraced by the iOS platform (both mobile Safari and the web view in Objective-C, which use the WebKit browser engine), we need to use the proprietary prefix **-webkit-**. Thus the CSS rule is **-webkit-gradient**, and our example is calling up a linear gradient so that is the next setting. Linear gradients can go from top to bottom, left to right, or even diagonally across the screen, so the next settings specify the start and end points as well as the direction. The last two settings are the start and end color; these can be hex, RGB, or named colors.

One more CSS adjustment will make this demo look great. Our current horizontal rule bleeds off the edge of the screen, but this might look a bit odd with the gradient added to the background. So let's shorten that up a bit with this addition to the style.css file:

```
hr {
  width:280px;
  margin-left: 0px;
}
```

To put this new style into action, we just need to modify main.html by adding the **.linear** style to the **body** element:

```
<body class="Juvelo linear">
```

Note that **linear** is appended to **Juvelo**—we need both classes because **Juvelo** is what styles the text on the page from the previous example.

Testing this revised code in Simulator will dress up the screen substantially with the new gradient in the background (**Figure 9.11**).

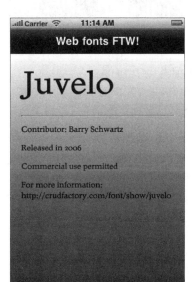

9.11 The Juvelo text screen with its new background gradient applied evenly from top to bottom.

What if you want the gradient to be diagonal? Simple: just change the direction of change from left top to left bottom, to left top to right bottom:

```
background: -webkit-gradient(
  linear,
  left top,
  right bottom,
  from(#fff),
  to(#333));
```

The result will look like **Figure 9.12.**

9.12 The Juvelo text screen with its new background gradient evenly applied diagonally from top left to bottom right.

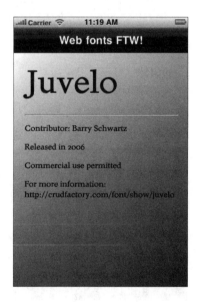

You're not required to have a gradient start at the top and end at the bottom; you can set the start and stop points from 0 to 1 so you get larger areas of solid color and a narrower band of the gradient. To see how this works, edit your code by adding **color-stop** to match this example, which constrains the gradient to the middle 40 percent of the page and allots the first and last 30 percent to solid color:

```
background: -webkit-gradient(
  linear,
  left top,
  right bottom,
  from(#fff),
```

```
color-stop (.3, #fff),
color-stop (.7, #333),
to(#333));
```

And now the results look like **Figure 9.13**.

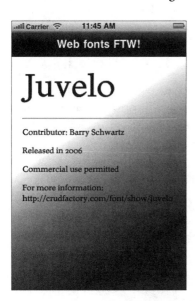

9.13 The Juvelo text screen with its new background gradient *unevenly* applied diagonally from top left to bottom right; the first and last 30 percent of the background is solid, and the gradient is applied only to the middle 40 percent of the distance.

Now the result is very different, and has more of a glassy ripple effect.

As you might have guessed, this is just the beginning of using gradients—there is also a radial gradient. Experimenting with gradients is even more fun when you transition from one color to another, so give that a try, too; I couldn't adequately convey the coolness of color gradients here in this two-color format. You're not limited to transitioning from just one color to another, either; you could incorporate several transitions in `color-stop` for a rainbowlike effect.

Cool CSS3 styling tools

Writing CSS3 is not particularly difficult, but perhaps you are more of a visual designer. Or maybe CSS3 is still really new to you and you'd like to start using it more, but keep the training wheels on for a while. Regardless of your reason, there are some really cool CSS3 tools out there that can help you work much more quickly, or more safely, as you explore your options. Two of them are AppControls, a shareware application, and CSS3 Please!, a cross-browser CSS3 rule generator.

AppControls

Like NimbleKit, AppControls is another iOS development tool that is featured on Apple's website—that's how I discovered it. In fact, it's not technically just an iOS development tool; you could use it to style controls for websites, too.

AppControls gives you a handy window where you can quickly specify width, height, border radius, gradient, border, drop shadow, inner shadow, and text options for controls. As you make your selections, the preview changes right in front of you (**Figure 9.14**).

9.14 AppControls provides an easy-to-use interface for making CSS3 selections, and provides a live preview of the results in the same window.

If the live preview isn't cool enough for you, the CSS that AppControls generates for you should be. That's right, click on Show CSS and you get all of the CSS you need to create the same look, either inline or via an external stylesheet (**Figure 9.15**).

```
#button {
    width: 198px;
    height: 51px;
    -webkit-border-radius: 12px;
    -moz-border-radius: 12px;
    border: 1px solid #5F5F5F;
    background: -webkit-gradient(linear, left top, left bottom, color-stop(0.00, #ADBDC8),
color-stop(1.00, #5F5E83));
    background: -moz-linear-gradient(top, #ADBDC8 0.00%, #5F5E83 100.00%);
    box-shadow: -0px 1px 3px #515151;
    -webkit-box-shadow: -0px 1px 3px #515151;
    -moz-box-shadow: -0px 1px 3px #515151;
    text-align: center;
    text-indent: 0px;
    color: #FFFFFF;
    font: 24px/51px "Helvetica";
}
```

| External CSS | Inline Style | Copy is disabled. Register to enable it. | Close |

9.15 AppControls also provides the CSS code, so it writes the button styles for you—and you can choose from the inline style or external stylesheet rules.

After you get the control looking the way you want to, just copy the CSS out of AppControls and put it into a stylesheet in an Xcode app. Reference the stylesheet in your HTML file, style a `div` with the ID that is provided, and enter the same text you used in AppControls—the result will look the same when you test it in Simulator (**Figure 9.16**).

9.16 AppControls also provides the CSS itself, so it writes the button styles for you—and you can choose the inline style or the external stylesheet rules.

AppControls is shareware that you can download from the folks at Blue Crowbar Software[11] and at this writing the full version costs just $29.90; the free trial version does everything for you, but it doesn't allow you to cut and paste the CSS that it generates (that is, some typing is required if you go the free route with this tool).

CSS3 Please!

CSS3 Please! is a website[12] that provides results that are similar to AppControls. The up side is that the website doesn't require you to sign up, so it's completely free. The down side is that the interface is less elegant, and doesn't allow you to change the content or size of the **div** that you're previewing. But for quick previews of CSS3 rules in a live browser setting, you still can't beat the price for what it can do.

The CSS3 rules that CSS3 Please! previews for you include border-radius, box-shadow, gradient, RGBA, transform (rotate), transition, text-shadow, and @font-face (**Figure 9.17**).

9.17 CSS3 Please! features a live preview of the CSS3 rules that you edit, but no UI is provided—you edit the code right in the same browser window!

Instead of color pickers and sliders, though, you just click on the CSS that is provided right on screen and make changes that are then displayed in the adjacent live view.

It's not quite as fancy as AppControls, but it's fun to give it a try just to see how they built the web tool. Kudos to its creators, Paul Irish and Jonathan Neal (in association with Boaz Sender and Zoltan Hawryluk), whom you should look up—they do some pretty cool stuff when they're not tweaking CSS3 Please!

These are just two of the many CSS3 tools and references out there. Keep searching for things that fit your app needs, personal preferences, and development budget.

Summary

HTML5 and CSS3 are still in the early stages of adoption. In fact, HTML5 is still at a point where the advantages are more behind-the-scenes and in the design phase of a project. Nonetheless, you have learned how to leverage several HTML5 and CSS3 rules for your iOS projects that are powered by NimbleKit. This gives you a leg up on the competition by providing you with applied HTML5 and CSS3 experience, which helps you not only design better iOS apps, but also better websites. Now it all comes full circle: iOS app design experience enriches your web design abilities and opportunities!

Next: NimbleKit might be cool, but are there other code frameworks for designing iOS apps without having to learn Objective-C? The answer is yes! The details follow.

REFERENCES

1. http://en.wikisource.org/wiki/Constitution_of_the_United_States_of_America

2. http://www.w3.org/TR/html4/struct/global.html#edef-DIV

3. John Allsopp, *Developing with Web Standards* (New Riders Press, 2009), p. 274

4. http://robertivan.com/WorldCuphtml5.html

5. http://vis.robbymacdonell.com/stanley-cup/

6. http://joshduck.com/periodic-table.html

7. More information about the CSS3 RGBA specification at http://www.w3.org/TR/css3-color/#rgba-color

8. http://www.w3.org/TR/css3-fonts/

9. http://kernest.com/

10. http://crudfactory.com/font/show/juvelo

11. http://bluecrowbar.com/

12. http://css3please.com/

10 OTHER MOBILE FRAMEWORKS

I will be perfectly honest: As cool as it is, the NimbleKit Objective-C framework may not be the right design or development tool for your particular app projects. I wrote this book because my mobile app design story is based on using NimbleKit, and the framework has worked extremely well for my projects and my collaborators' projects. I also know that I haven't even tried all of its features yet (and new features continue to be added), so I am confident that it will remain a robust platform for some time to come.

To me, this meant it was an important opportunity to tell other standards-based web designers about.

But you may have project requirements that NimbleKit doesn't meet or you may not wish to design only for iOS devices. And I'm not here to talk you into using NimbleKit exclusively—you definitely need to be savvy about using the right tool for the right task.

If this is the case, or if you're curious about exploring some additional options now that you're familiar with iOS app design, the mobile user experience, and the big picture of app design and planning from beginning to end, this chapter is for you.

Emulating the iOS experience with PhoneGap and jQTouch

Another framework option that I think is particularly compelling is PhoneGap, especially when used in conjunction with a styling and behavior plug-in such as jQTouch. This approach has some similarities to NimbleKit, but it also has some key differences.

Considering PhoneGap

PhoneGap is an open source framework that is free to download from its developer, Nitobi, Inc. (**Figure 10.1**).

10.1 The PhoneGap website.

This framework has a number of important things going for it. To start with, the fact that it is open source is a win for people who gravitate toward open source solutions. Open source solutions offer some distinct advantages:

- Downloading and upgrading is usually free.
- The continued development of the solution is less centralized, more organic, and tied to real-world needs as they crop up "in the wild," so this can make open source solutions feel more responsive to community needs.
- Some people just prefer to avoid corporate and proprietary solutions.

Remember that open source solutions can have down sides as well:

- Some argue that open source options are "free like a free puppy"—the solution itself may be free, but the ongoing support may have hidden costs, whether out-of-pocket or in terms of time spent finding free support.
- The development of open source solutions is dispersed throughout a community, but so is the responsibility for maintaining the tool and keeping it going (and decentralized responsibility can result in its own political challenges).

Another strength of PhoneGap is multiplatform support. It currently supports iPhone, Android, BlackBerry, Symbian, and Palm mobile operating systems. But the support and access to native features varies from platform to platform, and you still need to install (and learn how to use) the individual development environments for these platforms in order to use PhoneGap in that capacity.

So, what about the native iOS user experience? This aspect of PhoneGap is a mixed blessing, as the framework does not include native user interface library items such as a title bar, tab bar navigation, audio and video players, and other familiar iOS elements like NimbleKit does. From an interface design perspective, it is much more of a blank canvas.

But there are ways to emulate, or build, your own iOS interface, with CSS and JavaScript in PhoneGap. One way is by using it with the jQTouch plug-in.

Considering jQTouch

jQTouch is a jQuery plug-in for mobile web design on iOS devices, and can work in tandem with development frameworks such as PhoneGap to create native iOS applications. It can also be used for web apps that reside on servers and are delivered live to mobile browsers, on any mobile device (**Figure 10.2**).

10.2 The jQTouch website.

This jQuery plug-in has been optimized to run on mobile devices, and has native-looking styles that can help make your iOS app feel familiar to users of iOS apps. jQTouch accomplishes this by emulating native iOS interface elements and behaviors, such as page transitions, with convincing graphic files and CSS transitions.

How is this different from NimbleKit's library items? Because the NimbleKit framework is designed to serve only iOS devices, it includes prewritten Objective-C modules that call native iOS interface elements and behaviors. This is great if you want to design an iOS app that's as native as possible; the only emulation of iOS elements are those that you choose to add yourself. (An example is making buttons that are styled with CSS to be native-looking, which you explored earlier in this book.)

On the other hand, using jQTouch can lead to some pretty convincing results, too, even if the styles are only native-looking. Plus, if you want to design a less-native interface or behavior, using jQTouch with PhoneGap should afford you a wider range of possibilities to do so. Again, it's more of a blank canvas, just as an empty browser window is for a new website design.

In fact, to make this clear, jQTouch includes two predesigned styles for the plug-in. One is called apple, and the other is called jqt. The apple style does what you would expect, and the jqt style has a bit more of an Android feel to it.

For example, here's a version of the Bill of Rights app demo that you explored in Chapter 5, designed using PhoneGap and jQTouch with the style set to apple (**Figure 10.3**).

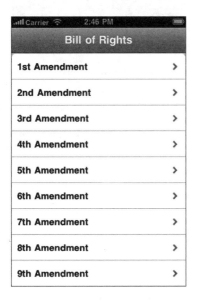

10.3 The Bill of Rights table view example using PhoneGap, jQTouch, and jQTouch's apple theme.

NOTE **More about PhoneGap and jQTouch** If you want to learn more about PhoneGap and jQTouch, I recommend reading *Building iPhone Apps with HTML, CSS, and JavaScript* by Jonathan Stark.[1] In fact, Jonathan has inherited the maintenance of jQTouch from its designer, David Kaneda. Learn more about Jonathan at his website, www.jonathanstark.com.

If you refer back to Figure 5.3, you will scarcely notice a difference in the graphic presentation between this native-looking example and the native interface of the NimbleKit example.

Here's the same screen using PhoneGap and jQTouch, but with the style set to the jqt theme instead (**Figure 10.4** on the next page).

10.4 The Bill of Rights table view example using PhoneGap, jQTouch, and jQTouch's jqt theme.

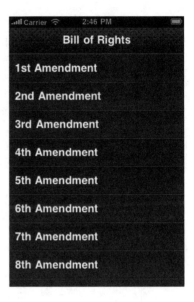

This is certainly a cool look. Is it a native iOS user interface? No, not entirely. But not all iOS apps need to stick to a strictly native user interface to be successful.

To give you an idea of how the HTML syntax differs from NimbleKit when using PhoneGap, here's the file for the apple-themed example shown in Figure 10.3:

```html
<html>

<head>
<title>Bill of Rights</title>
<link rel="stylesheet" href="jqtouch/jqtouch.css"
type="text/css" media="screen" title="no title"
charset="utf-8">
<link rel="stylesheet" href="themes/apple/theme.
css" type="text/css" media="screen" title="no title"
charset="utf-8">

<script type="text/javascript" src="phonegap.js"
charset="utf-8"></script>
<script type="text/javascript" src="jqtouch/jquery.js"
charset="utf-8"></script>
<script type="text/javascript" src="jqtouch/jqtouch.js"
charset="utf-8"></script>
```

```html
<script type="text/javascript" src="billofrights.js"
charset="utf-8"></script>
</head>

<body>
<div id="home">
  <div class="toolbar">
    <h1>Bill of Rights</h1>
  </div>
  <ul class="edgetoedge">
    <li class="arrow"><a href="#1">1st Amendment</a></li>
    <li class="arrow"><a href="#2">2nd Amendment</a></li>
    <li class="arrow"><a href="#3">3rd Amendment</a></li>
    <li class="arrow"><a href="#4">4th Amendment</a></li>
    <li class="arrow"><a href="#5">5th Amendment</a></li>
    <li class="arrow"><a href="#6">6th Amendment</a></li>
    <li class="arrow"><a href="#7">7th Amendment</a></li>
    <li class="arrow"><a href="#8">8th Amendment</a></li>
    <li class="arrow"><a href="#9">9th Amendment</a></li>
    <li class="arrow"><a href="#10">10th Amendment</a></li>
  </ul>
</div>

<div id="1">
  <div class="toolbar">
    <h1>1st Amendment</h1>
    <a class="button back" href="#">Back</a>
  </div>
  <p style="margin:10px">Congress shall make no law
respecting an
  establishment of religion, or prohibiting the free
exercise thereof; or
  abridging the freedom of speech, or of the press; or the
right of the
  people peaceably to assemble, and to petition the
Government for a
  redress of grievances.</p>
</div>

<div id="2">…
{amendments 2-10}
</div>
</body>
</html>
```

In addition to the slightly different HTML syntax (all of the screens are in one file), the highlighted links to CSS and JavaScript files show the most significant difference between NimbleKit and PhoneGap. But when using jQTouch with PhoneGap, as this example does, most of those files are provided for you in their respective downloads.

Developing native apps with Titanium Mobile

Another framework that supports both iOS and Android is Appcelerator's Titanium Mobile, and it's free to download from www.appcelerator.com (**Figure 10.5**).

10.5 Appcelerator's Titanium Mobile.

Titanium Mobile is a bit of an odd duck: It is an open source development framework with a very robust and glitzy marketing campaign behind it. So it is similar to PhoneGap in that it is a company product, but it has also been open-sourced. In this case, however, the presentation feels very corporate to me (but I guess it's because Appcelerator also sells support and training for Titanium Mobile).

I recommend that you check out Titanium not because I've used it (I have not), but because colleagues of mine have and they confirm that it is a good option. Here are some of the things they've said about Titanium Mobile:

- Although it has a large number of APIs (the website claims over 300), they are all accessible to the developer via web standards. But Titanium Mobile requires almost everything to be written in JavaScript. So on a spectrum of tools for designer-types versus tools for programmer-types, this one falls more firmly into the programmer camp.

- Titanium Mobile out-of-the-box comes with more corporate branding than PhoneGap or NimbleKit, which do not force any company branding into the interface. But some swapping out of images can get you around this (though I can't vouch for whether doing so voids any end-user license agreement clauses that you tacitly agree to when download-ing the tool).

Another key difference with Titanium Mobile is that the download comes with its own Mac OS desktop application, Titanium Developer (**Figure 10.6**).

10.6 Titanium Developer.

Titanium Developer manages, compiles, and packages project files. It effec-tively bypasses Xcode as your development environment, which means that you write all of your JavaScript in the editor of your choice. Once you're done writing the code and designing any other assets, you import the project files and compile, after which Titanium Developer launches Simulator (rather than Xcode) to show you the results (**Figure 10.7**).

10.7 Titanium Mobile's Kitchen Sink demo app in Simulator.[2]

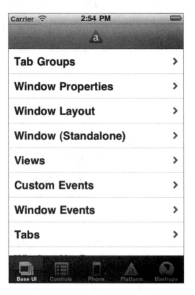

If you are an experienced JavaScript programmer and are interested in developing native apps for both iOS and Android, I would encourage you to investigate Titanium Mobile when evaluating tools for your app projects.

Designing web apps with Sencha Touch

Are native apps, designed for specific platforms and installed to run on devices, better than web apps? Or are web apps, which are more universally designed and run on servers just like websites, better than native apps?

Yes.

Although my focus in the mobile space has been on native apps for the iOS platform, I certainly would not discourage you from considering web apps to meet your app design needs or those of your clients. In fact, many of the ideas and skills you've learned in this book apply to designing web apps as well as designing native iOS apps.

When considering ways of designing effective mobile interfaces for universal web apps, you already have one option to consider: the jQTouch plug-in. Another option is Sencha Touch, which is free to download from www.sencha.com (**Figure 10.8**).

10.8 The Sencha Touch mobile app framework.

Here are some highlights of Sencha Touch:

- It prides itself on being the first app framework built specifically to leverage HTML5, CSS3, and JavaScript to maximize power, flexibility, and optimization. Web standards FTW!

- Sencha's beta version is compatible with both iOS and Android, with themes designed for both platforms. This is not technically a device platform limitation, however, but rather a WebKit rendering engine limitation. Other mobile browsers based on WebKit should support Sencha Touch quite well.

- Because the mobile web is evolving on most platforms to be touch-based, Sencha Touch has developed a robust range of touch interactions: tap, double-tap, swipe, tap-and-hold, pinch, and rotate. Such interactions that were initially only available in the native iPhone platform could eventually become ubiquitous on all mobile devices, with Sencha leading the way with its Touch web platform.

Remember that we are still very much in the Wild West of mobile app design, with these devices still being very new and the range of good design and development tools still growing. And when considering an app project for yourself, your employer, or a client, the NimbleKit Objective-C framework is a strong, solid option—but consider the full range of issues that go into selecting the right tool for the right job. **Table 10.1** shows how some of these issues compare across the tools mentioned in this chapter. It is by no means comprehensive, but serves as an example of how they can differ in both stark and subtle ways.

TABLE 10.1 Comparing NimbleKit to other mobile framework options

	NIMBLEKIT	PHONEGAP	TITANIUM MOBILE	SENCHA TOUCH
Native iOS UX	X	emulated	X	—
App Store distribution	X	X	X	—
Android	Fall 2010	X	X	X
Licensing	X	—	—	X
Web apps	—	X	—	X
Featured in Apple Developer Tools	X	—	—	—

Finally, note that VolnaTech, the developer of NimbleKit for iOS, released NimbleKit for Android in the fall of 2010. It promises to deliver a similar opportunity to web designers who want to design native Android apps by using web standards, for distribution in the Android App Store, without needing to learn Java programming or the Android SDK.[3] NimbleKit for Android is a separate license from NimbleKit for iOS.

Summary

In this chapter, you learned about three alternative methods for designing or developing mobile apps for iOS devices as well as other platforms:

- PhoneGap (with jQTouch) provides a web standards–based alternative that is similar to NimbleKit. It comes with fewer APIs than NimbleKit and emulates rather than calls native user controls, but also affords a wider range of UI styling options.

- Titanium Mobile is another web standards–based alternative, though it relies heavily on JavaScript programming and is thus more of a development framework than a design one. It comes with more APIs than NimbleKit, but uses its own desktop application for compiling and packaging the app binary.

- Sencha Touch is an HTML5, CSS3, and JavaScript web application framework that allows you to more quickly design mobile apps that run on web servers instead of being distributed or sold via the App Store. The product is not a native iOS app and lacks an iOS user experience, but still has a robust touch interface that is designed to work beautifully in WebKit-based mobile browsers on iOS, Android, and other mobile operating systems.

REFERENCES

1. http://oreilly.com/
   catalog/9780596805784?cmp=il-orm-ofps-iphoneapps

2. http://github.com/appcelerator/KitchenSink

3. http://www.nimblekit-android.com/

11 MARKETING YOUR APPS

I always tell people that I view design as a holistic activity: The visioning, planning, and creative work involves more than making the thing that you're designing. Much of the creative challenge lies in nurturing the human relationships involved in design projects, as well as managing the various outcomes that can occur after the "creative work" is done.

In other words, it's all design—and all of the work for a project, from beginning to end, is creative because it is all interconnected.

So as you travel the arc of iOS app planning, design, and management, you'll need to consider some of the more communications- and marketing-oriented decisions. Because creating an app is a lot of the work, but it's not all of the work. Who is the audience for your iOS app, who is the client, and how do you want to shape your relationship with these entities, and with Apple? And what are Apple's guidelines for doing this?

This chapter is less technical, but don't think that makes it any less important. Read on so you can be just as savvy with the communications and marketing aspects of your app as with the other creative aspects of iOS app design.

Who are you: Deciding on an App Store identity

This isn't just a rhetorical question. Seriously, who do you want to be when designing apps and making them available to customers? Here are some options to consider.

You are you

Perhaps you want to keep things straightforward and, if you already registered as an Apple Developer under your own name, do any app design consulting or collaboration with others under your own name. Similarly, you may want to submit apps to Apple as an individual as well.

Designing and submitting apps as an individual has the following advantages:

- You can distribute or sell your apps via the App Store.
- It is immediately obvious who has designed your apps because they're listed under your name in the App Store.
- There may not be any financial or legal hoops to jump through in your state to do business as an individual.
- It is likely to be faster and easier to do business as a sole proprietor, and submit apps to Apple as an individual.

On the other hand, designing and submitting apps as an individual may have the following disadvantages:

- You are probably personally liable for all business and design decisions that you make, in any event that an app client or customer has a problem with your work.
- You give up some anonymity and privacy by submitting and distributing iOS as an individual. After all, your name will be listed in the App Store!

- The annual cost of being in the regular Developer Program is currently $99. If you're registered as an individual developer, you're on the hook for covering the annual fee all by yourself (unless you recover the cost with consulting fees).

Take these considerations into account if you're thinking about operating as an individual or sole proprietor, and become familiar with local, state, and federal commercial and tax requirements that may pertain to your form of business.

You are your company

Perhaps you're like me and want to distribute your apps via a company name. When I formed my company (Aesthete Software, LLC), I did it for these advantages, so see if they fit what you are looking for:

- You want to look more established and professional by being listed in the App Store under a company name.
- You would like to limit your personal and family liability by creating a form of business, such as a limited liability corporation (LLC).
- You have identified tax advantages of doing work as a company rather than as an individual.

Of course, as with any option there are downsides to this one, too:

- It takes some time and, often, some expense to create a new legal entity for a company.
- There may be additional tax forms to complete—as if taxes were not enough work already!
- If you want to take this route to its logical conclusion, you could end up paying for additional company costs such as a new web domain, web hosting, and so on. (This could be in addition to any personal website expenses you already have.)

Even though I took this route for my app distribution, don't follow my lead without thinking through your entire business plan, budget, and willingness to try new things. Designing apps may be enough new work for you—creating and managing a new business entity may not be on your priority list at the moment, and that is just fine.

You are your employer or organization

Another option for designing and distributing iOS apps is to participate in Apple's Enterprise Program. The annual fee for this program is currently $299.

Designing and submitting apps as an enterprise has the following advantages:

- You should be able to get your company to cover the annual fee.
- You likely have in-house financial and legal resources available to you for consultation and assistance as you navigate the iTunes Connect business processes.
- If you are designing apps only for in-house use, the Enterprise Program allows you to distribute apps directly to employees in your company rather than going through the public App Store review and approval process.

On the other hand, designing and submitting apps as an enterprise may have the following disadvantages:

- You cannot sell or distribute your company's apps through the App Store.
- Your may get less personal recognition for your apps.
- Your company's apps may get less exposure (though if this route is taken for an internal app, additional exposure may not be relevant).

You are your client

Yet another option for designing and distributing iOS apps is to place them in the App Store not as yourself, your company, or your employer, but as your client.

Designing and submitting apps as your client has the following advantages:

- Your client gets to take on more of the copyright, liability, and financial issues of distributing apps via the App Store.
- This arrangement allows you to more comfortably collaborate on or design an app that doesn't fit into your overall market focus or app type specialization.

- It allows you to be a bit more risky with designing an app that might not sell or get reviewed well (though to be fair, your association with a bad app can probably come back to haunt you even if it's listed under your client instead of you personally).

On the other hand, designing and submitting apps as your client may have the following disadvantages:

- Depending on who administers the iTunes Connect account, you may not have financial control in this arrangement and will instead be paid by your client after Apple pays them.

- If the app ascends to a Top Ten list, you may regret not getting all the glory.

If you're like me, sometimes being confronted with too many choices can be unpleasantly bewildering, especially if you anticipated a more simple process. As in, "I just went to the store to buy toothpaste—I didn't expect to stand in the aisle, paralyzed by all the options!" This section is not meant to paralyze you, but I encourage you to weigh these options early in your planning process. My hope is that if you consider the options carefully and seek advice early and as needed, then you won't be found standing in the aisle with a finished app, wondering how to quickly choose an option that may be more than a five-minute decision.

Using Apple's marketing assets

Apple and its App Store have plenty of critics and, as I noted in the introduction to this book, my decision to write this does not mean that I dismiss all of those criticisms. The iOS app design and distribution processes have plenty of shortcomings, and there are plenty of reasons to perhaps be cynical as to why Apple has designed these opportunities and processes in the ways that it has.

Nonetheless, the numbers generated by the App Store are certainly some of the reasons to take interest in Apple's mobile platform and app distribution model. There are over 250,000 apps in the App Store, and by January 2010 more than 3 billion apps had been downloaded worldwide.

And that was before the iPad was introduced.

So how can you take advantage of this enormous opportunity and sales momentum for your app? One way is to take advantage of the marketing assets provided by Apple. But take care when you do—Apple's famous attention to detail and penchant for control means there are some rules about using them.

Using the Available on the App Store badge

The Available on the App Store badge has surely has become one of the most recognizable graphic icons in the world in just a few short years (**Figure 11.1**).

11.1 The Available on the App Store badge.

I have to admit, there was a brief yet huge moment of excitement when I added it to the website I had created for my first app—I had arrived!

Needless to say, you'll want to use this graphic on any communications about your app, whether you're distributing or selling, and whether you're doing it for yourself, your employer, or your client. The graphic is instantly recognizable and has become a potent symbol in the heated competition between mobile operating systems (and native versus web apps).

Note that when you use this graphic, you are subject to the guidelines, requirements, and limitations as defined in Apple's App Marketing Artwork License Agreement. Here are some of the more important guidelines for using the Available on the App Store badge:

- Do not modify the graphic in any way. This includes changing its color, rotating it, adding or subtracting any elements of the graphic, or rearranging any of the elements.

- Follow the clear space requirements for implementation: 10 millimeters around the graphic in print, and 40 pixels on the web.

- Create a clear hierarchy where the badge is secondary to your other app branding and identity elements.

- You are not limited to a white or black background as you might expect from Apple, but they do expect you to make a judgment call about the badge remaining legible and being clear of visual clutter or patterned backgrounds.

- You *must* link to your application on the App Store whenever the badge is used. To get your app's URL, find it in iTunes and select Copy Link in the small dropdown menu located under the app icon (**Figure 11.2**). Use this URL either as a link from your own website or as copy in other promotional materials.

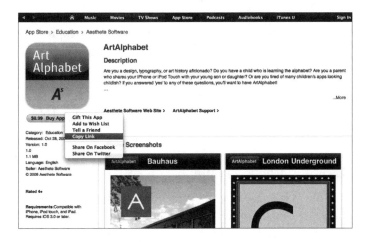

11.2 How to copy an app's URL in the App Store.

For additional details, refer to Apple's App Marketing and Identity Guidelines for Developers, available as a PDF in the iOS Dev Center.

Using official iOS device images

Once you agree to the App Marketing Artwork License Agreement and download the graphics provided by Apple, you are eligible to use the official iOS device images that are downloadable from the iOS Dev Center website.

As you have probably guessed, Apple has a handful of guidelines regarding your use of these images, too:

- Use only Apple-provided device images in your app's promotional materials. Do not take your own photos of your app running on a device, although taking your own screenshots is allowed and, in fact, necessary for showing your app. Instead, layer such screenshots on top of the device image provided by Apple.

- Do not use Apple iOS device images alongside an image of another mobile device. (This rule makes me laugh a bit!)

- Unlike the App Store badge, there is a strict background requirement for the device images: white is the only acceptable background.

- Do not place any promotional copy or images in the screen area of an iOS device image. The only acceptable images for the screen area are screenshots of your app as it runs on a device or in the Simulator.

- As with the App Store badge, do not modify device images in any way other than adding an app screen shot to the device's screen area. Prohibited modifications include changing color, rotating, adding or subtracting any elements of the graphic (including the iOS status bar), or rearranging any of the elements.

If you have any questions, please refer to Apple's App Marketing and Identity Guidelines for Developers for additional details.

Using Apple-approved language

Even when it comes to written descriptions of your app, Apple has a few opinions as to how you should proceed. For example, if your app is called AweSome MegaApp, you can refer to it as AweSome MegaApp for iPhone but not iPhone AweSome MegaApp in your communications about it.

Apple is also quite discerning about how you use the term App Store. Despite it being in iTunes, you are not allowed to add descriptors such as iTunes App Store or iPhone App Store. It's just App Store, and you are recommended to say things like your app being "available in the App Store" or able to be "downloaded from the App Store." And interestingly, Apple does not allow App Store to be translated into another language, even when used in communications in another language.

It is the App Store, and it shall remain the App Store.

Also, do not refer to Apple product names in your communications except for describing app compatibility either generally or specifically. And note that iPod touch is required to have a lower-case *t* in *touch*.

Did I mention that Apple was detail-oriented?

Some of Apple's language requirements are a bit tedious to describe, but I hope this is a good introduction to their top requirements. You are encouraged, however, to keep up with the details and any updated recommendations by checking in with the marketing resources on Apple's iOS Dev Center.

Using Apple-approved typography

Finally, don't think that Apple has no opinion on the typeface you use in your app communications—it does.

Fortunately, the rule is quite simple: You are forbidden to use any version of the Myriad typeface, because Apple uses a version of it in their corporate communications and they do not want your communications to be confused with theirs. Fair enough, though too bad for us (as Myriad is a nice typeface).

Designing your own app marketing communications

The fact that the App Store has over 250,000 apps means that it has a ton of traffic, and is currently the main point of contact for the majority of people in the world who are shopping for native mobile apps. No other mobile device has an app store with as many products or customers. This is great news for iOS app designers.

But having over 250,000 apps also means that your app is the proverbial needle in the haystack. Depending on what it does, your app could have a lot of competition. And depending on how it is categorized and positioned in the App Store, it could have either thousands of downloads or just a handful. This is the downside of the App Store for iOS app designers.

Naturally, success is defined by the scope and goals of any given app. A very specialized app might be designed to appeal to only a very narrow audience; most of the ones I have designed fall into this category. Nonetheless, I still want my target audience to find it. Using the web is a valuable tool in helping to get the word out.

So whether you are seeking exposure to all iOS device owners for a general purpose app or one that could have very broad appeal, or targeting a specific profession or audience with a niche application, consider your online communication options during your design process.

Designing an app website

Apple requires a URL for any app that is submitted to the App Store for approval. This URL could be a blog, a personal website, your employer's website, or your own company website—but it needs to go to a real website.

And let's face it, any additional publicity for your app is good publicity. An app website (or page) helps your app feel "more real," it may be able to inform people more quickly about what it does, and it gives you a much wider range of options than relying on the App Store alone.

I won't pretend to be an app marketing guru, although I did work for several years in marketing (but that was a while ago and not in this industry). Still, I am happy to share some recommendations based mostly on my own experience with my company's website, www.aesthetesoftware.com (**Figure 11.3**).

11.3 The Aesthete Software, LLC, website.

Here are some decisions I made concerning this website and how it relates to marketing the apps I have designed:

- The branding of Aesthete Software, LLC, has been very consistent. The details of the wordmark (typeface, color, spatial relationships) have been consistent between the website and the app graphics in the App Store, as well as in the apps themselves.

- I do my best to maintain an attention-getting home page, usually either with fresh news or with a large image and quick message.

- The navigation for this site has been kept quite simple. I have sections for apps, consulting, workshops, and now this book.

- To keep updating simple and easy to access, I decided to use a web-based content management system. This allows me to add or revise content wherever I am as long as I have an internet connection.

- I incorporated a Twitter feed in my site's design, so that even if the site's home page has not been updated recently (which I can be guilty of when busy with other things), there is still current content streamed to the right column of the page (assuming I've kept up on Twitter).

In short, I encourage you to be somewhat Apple-like in your approach to communicating about any apps that you design. Use consistency to your advantage: You have not only created an app, you have created a new brand. Read up a bit on branding so you understand what it means to "manage a brand" because, after all, you or someone you are working with will need to skillfully and consistently stay on message with what your app does and for whom.

And if you are creating a suite of apps, be sure to plan how you will tie them together visually and textually. The relationships between you or your company, your app names, and your app graphics will either reinforce or detract from a sense of cohesion in your communications.

Creating an app social media channel

These days, the App Store and a website are just two parts of the puzzle. Social networks are carrying a large amount of today's web communications—and I'm sure you don't need me telling you this. You've probably logged into Facebook or Twitter today already or may even be logged in at this very moment, as you're reading this with another part of your over-active brain. So PAY ATTENTION: I have a few words to say about social networking on behalf of your app.

The most important thing to remember about creating a Facebook page, Twitter account, or other social media channel for your app(s) is that creating it is the easy part. Consider the Twitter account for my apps (**Figure 11.4**).

11.4 The Twitter account for Aesthete Software, LLC.

I'll be the first to admit that my best advice for social media is "do what I say, not what I do." Although I have been very active on this channel at times, I have also allowed it to languish at other times.

That's because unlike creating a social media channel, maintaining one and doing it well is extremely difficult.

The difficulty is not particularly technical, but rather it involves discipline. If content feels old on a website in a matter of days or weeks, content can feel old in a social media channel in a matter of hours.

And what will your app updates be about? New content? Version updates? Good. What kind of muffin you are having with your tall skinny cappuccino? Maybe not. Remember that if you open up a new social channel for your app, your content better stay app-focused for the most part. Too much filler will look like you didn't need the social channel in the first place.

Furthermore, I have learned that very few of my apps' users are even active on social networks, at least when it comes to their work. So try to get a sense in advance about your audience's use of social media before embarking on this route for your app (and that, of course, could be tricky and take some time).

There are so many recommendations out there for how to adopt social media to your advantage, I don't want to rehash it all here. But I would encourage you to consider social media for your app or apps, yet caution you not to think of it as a silver bullet to success. Adopting social media is like marathon training or wine making: It inherently takes time and cannot be rushed, because you need time to build an audience, a social persona, and a pattern of communications that your audience can learn to expect and rely on.

So by all means consider it, but consider it with care.

Communicating via app updates

You may not have thought much about them, but the update messages that you get in the App Store app from designers are also a form of app marketing communications. But the content and format can really vary, and sometimes you are just as likely to skip these messages. What makes them work well, and what makes them work not so well?

For better or worse, these update messages are the most direct way you have of communicating with your app's customers after they've made their free download or purchase. **Figure 11.5** (on the next page) shows one example of an update message.

11.5 An update message for Lose It! in the App Store, as viewed on an iPhone.

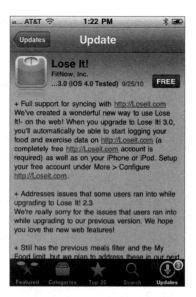

Here are some of the strengths of this message:

- It has a friendly, conversational tone, like the update is speaking with you.
- It is detailed and thorough.
- There are URLs included that link to pages on the app's website.

And here are some of the weaknesses of this message:

- It is really long.
- The bullet points are quite verbose.

As a comparison, **Figure 11.6** shows another app update message for Pandora.

The strengths of this example are clear: The bullet points are short and sweet, and not too numerous, so they are sure to be read. On the other hand, less than 30 percent of the viewable screen is used (not to mention that the update screens scroll, so they can continue below "the fold"). Was this a missed opportunity to do a bit more relationship building with Pandora's customers?

11.6 An update message for Pandora in the App Store, as viewed on an iPhone.

I think so.

Here are my recommendations for using app updates messages:

- Learn from these examples and aim for the middle of these two extremes. Be friendly and detailed, but also be concise. Keep individual points short, too.

- Always use a character as a bullet for bullet points, and always put a space in between them. If you don't do these things, your points are lost when they run together into one giant, gelatinous block of text that people will ignore.

- Try to keep your update message within the confines of the iPhone screen. This is about 18 lines of text (at approximately 50 characters per line). And don't be tempted to make your update messages much longer for an iPad app just because the screen is larger.

- Use updates as an excuse to remind customers to email you with feedback or suggestions, or rate your app in the App Store.

That is, make it more like the example shown in **Figure 11.7.**

11.7 An update message for TripIt in the App Store, as viewed on an iPhone.

Be creative with this rare opportunity to communicate with your customers and market new or updated features to them. If you can be concise, informative, and friendly, your app updates will enhance your and your app's reputation and help your customers remember to refer your work to others.

Summary

In this chapter you learned how to

- Consider what kind of presence you want in the App Store: yourself, your company, your employer or organization, or your client.

- Use the Available on the App Store badge and device images provided by Apple (and use them properly).

- Follow Apple's text guidelines for marketing messages, and how to *not* typeset your text in Myriad.

- Create your own app website and social media channels that communicate clearly to your customers, and augment the online presence of your app beyond the App Store itself.

- Write app update messages that are clear, concise, readable, and helpful—and strengthen your relationship with your customers.

You've now traveled nearly the entire iOS app design journey: from getting the app design tools you need, to learning about the iOS interface and user experience, to trying out techniques for designing various kinds of app content. This, coupled with marketing guidelines and tips, means that you're ready for the finish line: distributing your app via the iTunes App Store.

12 PROVISIONING AND DISTRIBUTING YOUR APPS

This journey started with the big picture of iOS app design, moved into the primary tools and methods for app design work, and has spent the last several chapters detailing various content-focused approaches to designing apps with web standards and NimbleKit's Objective-C code framework.

So what do you do when you are about to complete an app and start thinking about testing, distributing, and selling it?

That's what this chapter is about. One of the major reasons to "go native" and design an Objective-C iOS app is Apple's iTunes software and distribution channel. iTunes Connect is your interface with Apple's digital marketplace, and the iOS Dev Center is where you'll find the technical resources you need to make iTunes Connect work on your behalf.

Using the iOS Dev Center

The iOS Dev Center has added a lot of new technical content since the introduction of the iPad and iPhone 4 (and has always had a lot of information about the iPod touch). The Dev Center is a large, sprawling portal of information, resources, and forums for supporting your work as an app designer. This book will not attempt to catalog everything that's there— that would take several volumes, and it's actually just as effective and fun to poke around the place much like you would at any other large emporium. Rather, I'll focus on two key areas: the iOS Provisioning Portal and iTunes Connect.

Using the iOS Provisioning Portal

The iOS Provisioning Portal is the place where you can provision an app to *any* of the iOS devices you own or have access to. I'll walk you through the process, as it is somewhat tedious at times.

The reason that the app provisioning and distribution process is so tedious can be summed in one word: security. At least, this is what I've been told and this is what I choose to believe to avoid going crazy as I deal with all the steps that are required.

But it makes a lot of sense, as the system is designed around a set of digital keys, or certificates, that identify each essential component involved in the process: you, your app, and your device(s). These certificates and the processes that they plug into would make it extraordinarily difficult for someone to test or distribute your apps without your permission. Given how easy it is to copy digital files, we need to be thankful for this security.

On the other hand, the processes require either a certain amount of experimentation (and often debugging) to work properly, or some easy-to-use documentation. You probably don't want to do any more of the former than you really have to, and documentation is found in abundance in the iOS Dev Center—but it's not always filed under "easy" (in fact, it's not always even up-to-date).

That's where this chapter steps in.

Requesting and installing certificates

In order to run on an Apple mobile device (that is, on anything besides Simulator), an iOS app must be signed by a valid certificate that is obtained from the iOS Provisioning Portal. To run on your own device (or any that you have direct access to), you'll need a *development certificate*. For customers to run it on their devices after obtaining it from the iTunes App Store, a *distribution certificate* is required.

The first step in this process is generating a Certificate Signing Request (CSR) in the Keychain Access application on your Mac. If you're not familiar with this app, it is in Applications > Utilities.

After you start Keychain Access, go to the main application menu and select Preferences. Click on the Certificates tab and be sure that the first two options are set to Off (**Figure 12.1**).

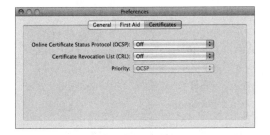

12.1 The Certificates Preferences in Keychain Access.

After you verify these preferences, go back to the main application menu and select Certificate Assistant and then choose Request a Certificate from a Certificate Authority (**Figure 12.2**).

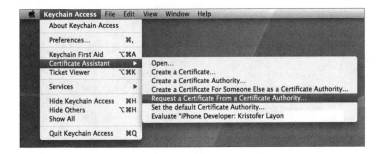

12.2 Request a Certificate from a Certificate Authority.

Although you need to have an email address entered in the top field, email address fields here are not actually used for this particular process. But you do want to be sure to use the exact same name you used when registering for your Apple Developer ID. You also want to select Saved to disk as well as Let me specify key pair information (**Figure 12.3**).

12.3 The Certificate Assistant.

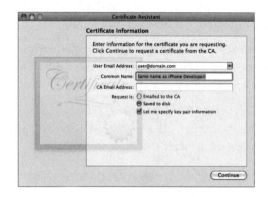

NOTE

Do I have any idea what these settings mean? No, none at all—but you can read more about them in the Dev Center if you really want to.

You will be prompted to save the Certificate Signing Request (pay attention to where you are saving it!), and then you'll see the screen shown in **Figure 12.4**. Be sure it looks just like this, with Key Size equal to 2048 bits and Algorithm set to RSA. Click Continue.

12.4 Saving your CSR and reviewing Key Size and Algorithm settings.

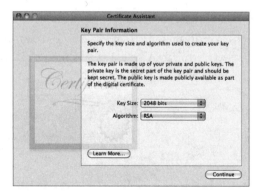

After you save your file, you will be the proud owner of a unique item: the smallest file (4 KB) with the longest file extension (.certsigningrequest) you're likely ever to have on your computer. Then you can log in to the iOS Provisioning Portal, go to Certificates and Development, and click Add Certificate.

NOTE A note regarding the distribution certificate

To request and generate a distribution certificate, it's almost the same process. Repeat these same steps but go to the Distribution tab under Certificates instead.

Once your development certificate is approved, you will see results like this (**Figure 12.5**) indicating that you may download the certificate.

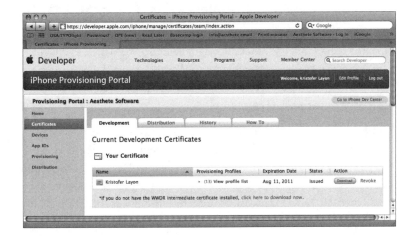

12.5 Congratulations— you have a development certificate!

Once you download the file, just drag the certificate to your Keychain Access app icon to install.

After you've done this, you should be able to create and download app provisioning certificates (described in the next section). But you need to repeat the above process once you wish to request, download, and install a distribution certificate, which you will need to distribute your app to customers via the iTunes App Store. After you have both your provisioning and distribution certificates, your Keychain Access should look something like this (**Figure 12.6** on the next page) when viewing your certificates.

12.6 Provisioning and development certificates in Keychain Access.

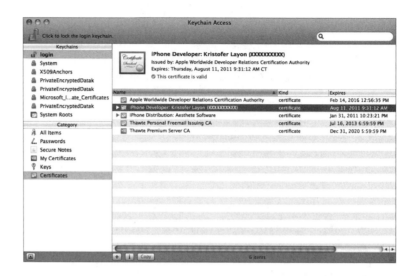

Note the presence of one other important item that you'll need to download from the iOS Dev Center: the Apple Worldwide Developer Relations Certification Authority certificate (referred to in some documentation as the WWDR intermediate certificate).

Finally, note that these certificates can occasionally be a pain to manage. For example, I ran into problems when my development certificate expired (they are valid for one year; this is tied to the regular Developer Program having an annual fee). It was easy enough to request a new one, download it, and install it. But much to my surprise, the old one was still in Keychain Access and interfered with the new one being recognized. I had to hunt around in Keychain Access, looking through Certificates and My Certificates until I found it and deleted it.

So beware: Deleting a certificate in one place does not always delete it completely from your system. Check both My Certificates and Certificates if you ever have to troubleshoot certificate issues.

Using the Development Provisioning Assistant

I have to take a moment to thank Apple for an improved process here, because the Development Provisioning Assistant (**Figure 12.7**) is a nice addition to the Dev Center. It does a much better job of helping you get from testing a new app in Simulator to testing on a device, and then eventually distributing via iTunes Connect.

This section explains how it works.

12.7 The Development Provisioning Assistant.

Creating an App ID

The initial view of the Assistant is rather nice and friendly (you're welcomed to it, in fact!). As I noted above, the screen then explains how some digital certificates are needed to match the right developer with the right app, as well as the right device. We just covered the development certificate process, so next we can create an App ID. The App ID itself is not a certificate that you download, it is just a name and character string used by the Dev Center and, eventually, iTunes Connect and the iTunes App Store, to identify your app.

When prompted by the Assistant, select Create a New App ID and you will get the screen shown in **Figure 12.8**.

12.8 Creating a new App ID.

Be sure to just give this a concise ID for your own purposes—this name is not particularly important. For example, some of the apps I have developed have longer names, but I tend to give them an ID that is abbreviated; only alphanumeric characters and spaces are allowed.

Choosing an Apple device

After you create a new App ID, you have two of the three digital components required to form an app provisioning profile (the development certificate and an App ID). It's kind of like an adventure story where you have to collect the missing objects that, once together, fuse into an orb that gives you magic powers.

And, as you're noticing, this process is nearly as adventurous!

When prompted by the Assistant to choose a device, select Assign a New Apple Device and you'll get the screen shown in **Figure 12.9**.

12.9 Creating a new device description.

The description itself should just be a short and simple phrase ("My iPhone 4" works), and then you need to get your device ID and enter it into the form. There are two ways to find the ID:

1. Connect your device to iTunes and click on the serial number in the Summary screen. This will display the device ID, though here it is called the Identifier (UDID).

2. Connect your device to Xcode and look for the Identifier in the Organizer window under Devices > Summary. It's actually a bit easier to get this way, as Xcode lets you select it to copy. (iTunes does not.)

After you finish assigning your device, the Assistant will verify the development certificate we created earlier.

Naming, generating, and downloading the provisioning profile

Now the Assistant displays the three magic items that, when fused, will give you magical superpowers—or, at least, the ability to install and test your app on a device (**Figure 12.10**).

12.10 Viewing your App ID, device, and certificate name.

It then should display a green check mark and verify that it has created the provisioning profile. Click Continue to download.

Next you're presented with a screen where you have to give the profile a description. Again, be brief and descriptive as you will be referring to this in Xcode settings—over time, you may have several apps and devices to test them on. I always use a combination of things, such as "App1 iPhone dev." This helps me distinguish my profiles according to app name, device, and whether the profile is for development or distribution.

After you give your profile a name, continue to follow the screens and prompts to download the profile and install into Xcode. (Dragging the downloaded file, which will end with the file extension .mobileprovision, to the Xcode icon is the easiest way to do this after you download it.)

Now you're ready to install and test! That process is detailed in the Provisioning and testing on a device (debugging) section in Chapter 3.

Using iTunes Connect

NOTE Distribution certificate

Remember that you can proceed with the following steps only if you have requested, downloaded, and installed your distribution certificate. If you haven't, go back to the Requesting and Installing Certificates section at the beginning of this chapter.

After testing and fine-tuning your app on a device or two (the more, the merrier), and with all of your other ducks in a row (some of which we will cover in this section), you are probably eager to get your app out into the wider world.

This means you're ready to distribute. You are also ready to start using iTunes Connect—almost.

Goodbye provisioning assistant, hello (again) Provisioning Portal

There is no distribution provisioning assistant like there is for app development, and I suspect it's because most of the steps are already completed so it's not really necessary. After all, you already have a development certificate and App ID, and distribution to iTunes does not involve a particular device ID like development testing does. So the last step is to generate a distribution provisioning profile.

To do this, go back to the iOS Dev Center and the iOS Provisioning Portal. Click on Provisioning and then click on the Distribution tab. Select New Profile and you will see a screen like what's shown in **Figure 12.11**.

12.11 Creating an iPhone distribution provisioning profile for your app.

(Pardon Apple's insistence in this corner of the web that you're only working on iPhone projects—the distribution provisioning profile works for iPod touch and iPad apps, too!)

Select App Store for the distribution method; give the profile a name similar to your development profile for the app. Again, I use names that include "Dist" at the end to help me keep my profiles separate when I see them in the Xcode organizer. You see your distribution certificate listed under this row, then all you need to do is select the matching App ID that you created during development provisioning. After clicking submit, you will return to the home view of the Distribution Provisioning tab and see your new distribution provisioning profile listed.

The rightmost column for your new profile will be blank at first. But by the time you reload the page, there will be a download button. Click it to download the file, and once you drag your freshly minted distribution provisioning profile to Xcode it should show up in your Organizer and you're ready to package and distribute your app binary.

This is detailed in the Building and submitting (distributing) section in Chapter 3.

Adding and managing applications

Now is the moment to break out the band: After you generate your app binary (as described in Chapter 3), it's time to submit your app to Apple for review and placement into the iTunes App Store.

Yes, the moment has finally arrived!

To start this process, go to itunesconnect.apple.com and log in using your Apple Developer ID, the same ID you use for the iOS Dev Center. You will be presented with a screen that has several options including Sales and Trends (which will become your favorite place to visit when people start finding your app); Contracts, Tax, and Banking Information (there are several important items to complete here prior to submitting your first app); Financial Reports (if you decide to sell your app instead of distributing for free, this will be your other favorite place to visit, because it documents Apple's payments to you), and some other categories for managing other users, in app purchases, and promotional codes.

After completing the steps in Contracts, Tax, and Banking Information, you begin the app submission process by clicking on Manage Your Applications. Then click on the blue Add New Application button in the upper-left corner of the screen.

The first screen you will see is the App Information screen (**Figure 12.12**).

12.12 Providing information about your app.

The first field, App Name, is the name that will be displayed in the iTunes App Store. *This name does not need to be identical to your actual app binary,* though it should be very close. For example, if your app name as installed is "Hello World" and your company name is da Vinci Design, Inc., you may want to capitalize on your ancestral relationship to Leonardo da Vinci and give your app the name "Hello World (da Vinci)" so that when it shows up in iTunes search results, the "da Vinci" grabs people's attention.

But except for minor alterations like this, the App Name should, in fact, be your app's name.

The next field, SKU Number, can be whatever you wish it to be. If you will be working on apps for others, and they have a particular system for their company's SKU numbers, you will want this number to fit into their merchandising system. But aside from these customer-side requirements, Apple has no particular guidelines for this field.

The last item in App Information is the Bundle ID. Use the dropdown to select the bundle ID that you created when first using the Development Provisioning Assistant to create your app's provisioning profile. And note that this ID should look familiar already because it matches the bundle ID field in your app's Info.plist file. (If it does not, iTunes Connect will not accept your binary.)

After you click Continue, the next screen presents you with the Rights and Pricing screen shown in **Figure 12.13**.

12.13 Determining rights and pricing for your app.

App approval time varies a lot (for me it has taken as little as two days and as much as three weeks). The important thing to know here is that Availability Date can be left as it is preset (to today's date), which means that your app will go on sale as soon as it is approved because that date will obviously be after this availability date of today. Or, if you know that there are a bunch of corporate marketing and PR reasons for launching a client's app on a particular date in the future, you can set the date accordingly.

Price Tier is labeled a bit strangely because Tier 1 is really just $1 (or, technically, $0.99). The tiers in the dropdown list go to Tier 85, and I hope that if you are submitting an app to sell for $84.99 and it actually moves at this price point, you will be sending me a small cut of the proceeds for getting you there.

And what if you have an app to sell for more than that? (Yes, they are out there.) Apparently you need to give Apple a call for some special treatment—I don't know how to submit an app for a price tier higher than what is in the dropdown menu.

The Discount for Educational Institutions checkbox enables schools, colleges, and universities to purchase your app in bulk at a 50 percent discount. If you're interested in this opportunity, read more about it on Apple's website and check this box when submitting your app.

Upon completion of this screen, click Continue and the Version Information screen appears (**Figure 12.14** on the next page).

12.14 Providing version information for your app.

Version Information

Enter the following information in **English**.

Metadata

Version Number

Description

Primary Category   Select

Secondary Category (optional)   Select

Keywords

Copyright

Contact Email Address

Support URL   http://

App URL (optional)   http://

Review Notes (optional)

Rating

For each content description, choose the level of frequency that best describes your app.

App Rating Details ▶

Apps must not contain any obscene, pornographic, offensive or defamatory content or materials of any kind (text, graphics, images, photographs, etc.), or other content or materials that in Apple's reasonable judgment may be found objectionable.

Apple Content Descriptions	None	Infrequent/Mild	Frequent/Intense
Cartoon or Fantasy Violence	○	○	○
Realistic Violence	○	○	○
Sexual Content or Nudity	○	○	○
Profanity or Crude Humor	○	○	○
Alcohol, Tobacco, or Drug Use or References	○	○	○
Mature/Suggestive Themes	○	○	○
Simulated Gambling	○	○	○
Horror/Fear Themes	○	○	○
Prolonged Graphic or Sadistic Realistic Violence	○	○	○
Graphic Sexual Content and Nudity	○	○	○

EULA

If you want to provide your own End User License Agreement (EULA), click here. If you provide a EULA, it must meet these minimum terms. If you do not provide a EULA, the standard EULA will apply to your app.

Images

Large 512x512 Icon ?

Choose File

iPhone and iPod touch Screenshots ?

Choose File

iPad Screenshots ?

Choose File

This screen has a lot of information that you'll need to answer very thoroughly. Some items like version number, copyright, email address, and support URL should be quite straightforward for you to answer. Similarly, the ratings section in the middle is also fairly objective.

But some of these items are of paramount importance to the positioning and marketing of your app in the iTunes App Store. Consider the description: How can you engage your prospective customers with a few paragraphs that are both attention-getting yet concise? What are the key features to describe, and if your app has competition, how can you distinguish your app from the others that are similar?

NOTE Version information—now editable

An important detail to note, however, is that you can edit the description and other items later. This is a new feature; before the current version of iTunes Connect, the only opportunity you had to change an app description was when you submitted an updated version for approval!

Equally important as the description, and perhaps even more so, are the categories and keywords. After all, people searching for your app by these broader terms will not even find it if you don't select these carefully.

The available categories are shown in **Figure 12.15**.

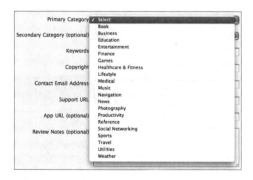

12.15 Determining the categories for your app.

Note that there is a primary category as well as an optional secondary category (the options are the same under both). I recommend using the secondary category, in particular, when your app does not comfortably fit into just one—at least when using it, you can broaden your reach with a second category!

Keywords are similarly important, and here is another key thing to note: App descriptions are apparently not used by iTunes search (at least, this is what Apple engineering staff said at the 2010 Worldwide Developer Conference). So this means that even if important keywords are used in your description, they also need to be entered into the keywords field so they are searchable by iTunes App Store shoppers.

NOTE   Version information—some is editable, some is not

An important detail to note, however, is that you can edit the description and latest update information later. However, you *cannot* edit categories or keywords later! Can. *NOT*. Choose these *very* carefully, as they are not editable until you submit a new version of the app for Apple's approval.

After completing the categories and keywords, you are likely to accept the standard End User License Agreement (EULA) that is provided by Apple unless you wish to retain a lawyer and write your own (to each his or her own). I don't want to suggest that Apple's EULA is perfect and cannot be improved, but I've been using it as is for my own applications.

The final task to complete is uploading the iTunes App Store icon graphic and app screen shots. To keep things consistent, I recommend generating PNG images for these just as you did for the app icons themselves; but note that iTunes also accepts JPG and TIF formats as long as they are 72 dpi, RGBA, flattened, and without transparency. Also, their file names do not matter as they do for app icons. **Table 12.1** shows the size specifications.

TABLE 12.1 Size specifications for iTunes app icon and screen shots

	APP ICON	IPHONE/ IPOD TOUCH	IPHONE 4	IPAD
Portrait	512 × 512 px	320 × 460 px	640 × 920 px	768 × 1004 px
Landscape	—	480 × 300 px	960 × 600 px	1024 × 748 px

The screenshot sizes in Table 12.1 assume that you are cropping out the status bar, which is 20 pixels high on the iPhone, iPod touch, and iPad (40 pixels high on the iPhone 4's Retina Display). I recommend this because I find the status bar to be distracting in screenshots: Nearly every app displays it, so it's not a distinguishing feature. But I've also mistakenly uploaded screenshots with it included and that's not the end of the world. Just realize that when you do this, the second dimension is larger than shown in Table 12.1 (e.g., iPhone portrait: 320 × 480 pixels).

Also, both portrait and landscape screenshots are displayed in the desktop version of the iTunes App Store on desktop and laptop Macs. To view them in the App Store on an iPhone or iPod touch, just rotate the device to landscape. In other words, they are not resized to display in portrait orientation on those devices.

After submitting the version information, you get a summary screen of your information. At this point you can either leave and come back, or wait a bit and refresh the view to disclose a blue Ready to Load Binary button. After being asked to answer a few questions, you'll be prompted to use the Application Loader. If you have downloaded iOS SDK 3.2 or later, you already have Application Loader stored on your Mac (in Developer > Applications > Utilities). When you run it, you will be prompted to Choose an Application (**Figure 12.16**).

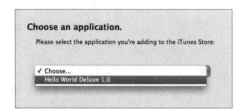

12.16 Using Application Loader, a small helper application that is part of the iOS SDK on your Mac.

Then you are asked whether you tested and qualified this binary on iOS4 (if you're using the current SDK and tested in the iPhone 4 Simulator and/or on an iPhone 4, answer "yes"). Once you select the app binary that you built in Xcode (it's in the build > Distribution-iphoneos directory of your main app file directory, and should be compressed so that it's in the format *name*.app.zip), Application Loader will submit it to iTunes Connect for you.

That's it—you're done! Now all you have to do is keep logging in to iTunes Connect (or check the iTunes App Store) to see when your app is approved and available.

Summary

Whew! This chapter basically gave you a master's degree in Apple Developer security and app marketing. I hope this documentation fills in a lot of the gaps that I found when I learned these processes and read Apple's documentation—if nothing else, I think it's helpful to have it all in one place. And what you just learned is how to

- Generate Certificate Signing Requests (CSRs) using the Keychain Access and Certificate Assistant applications on your Mac.

- Pass along the CSRs to the iOS Provisioning Portal to create development and distribution certificates.

- Download and install these certificates, along with the Apple Worldwide Developer Relations Certification Authority certificate (WWDR intermediate), in Keychain Access.

- Create App IDs and choose Apple Devices for development provisioning.

- Submit app information, version information, and app icons, screenshots, and binary files to iTunes Connect.

You've made it! You have explored how to design native iOS apps with web standards and NimbleKit, plus you have learned about some helpful UI and UX recommendations, a generous helping of HTML5 and CSS3, and some NimbleKit alternatives for mobile design.

And before diving right into your own projects, read about content strategy, planning, and usability (Appendix A). These topics did not fit neatly into the flow of the other chapters, but are perfect to read now before you embark on your own app adventures.

APPENDIX A

ADDITIONAL GUIDING PRINCIPLES

What follows are three additional sections about topics that are critical to designing successful projects: content strategy, application planning, and application usability.

These were originally conceived as shorter sections that were to be sprinkled throughout the book, flavoring the soup, so to speak, with broad and unifying principles that you should keep in mind from beginning to end.

The trouble is, they're all critically and equally important. But the rest of this book has a flow from preparing, to building, to the delivery of an app. So interspersing these extra sections throughout the book would have given a false impression that whichever one was last could be thought about toward the end of a project.

So this appendix is a bit of a paradox: It's at the end of the book, but you could (and even should) read it first. And you should certainly keep it in mind from the beginning of an app project to the end of its design and delivery, and beyond to include the app's ongoing maintenance.

In a sense, having this at the end is appropriate. Good content strategy, planning, and usability are the be-all and end-all of good design.

So here's the final word.

Content strategy

Content. We web designers are often at a loss about how to handle content. Some quick scenarios remind me of this:

- The *I-don't-really-care-about-content designer*: This person is all about deferring the responsibility of content to someone else. Heck, anyone else. Because we have plenty to focus on already, thank you very much: information hierarchy (OK, so that vaguely involves content), navigation, site structure, page layout, and the famously ambiguous but all-important look and feel.

- *Mr. or Ms. Lorum Ipsum*: Here's a designer who is kind of good at feigning interest in content. But is it really a sincere interest if the words are fake?

- The *I'm-a-Renaissance-man (or woman) designer*: This person is pretty confident that she can just massage the kinks out of any content that's handed to her. How hard can it be, especially with spell and grammar check turned on?

The trouble with having any of these attitudes regarding web content is that they allow us to dodge our overall responsibility for a high-quality, user-focused, and organization-focused website. Sure, we like to think that we're just focusing on what we should—design—but in reality, we're not fully focusing on design if we're being cavalier about content.

And the trouble with bringing these attitudes to the design of iOS apps is that it's even more dangerous than with websites. As you've already learned, there's a lot less space or interface to design in a mobile app. You simply don't have the luxury of saying, "Well, I have all of this design work to do—you take care of the content and tell me when it's ready."

Unless your app is a game, your mobile app **is content.**

Which means you need a content strategy.

Fortunately, content strategy is now seeing the light of day as a critical part of design projects. And one of the people who is shining a lot of light in this area is Kristina Halvorson, who gave us a wonderful book called *Content Strategy for the Web*.[1] If there's a recent book about content-focused design that's a must-read, this is it. Meanwhile, here's how some of its lessons can help us with designing iOS apps that are distributed in the App Store.

Less is more

So you thought Ludwig Mies van der Rohe, the great German Modernist architect, was the person who popularized the maxim "less is more"?

Well, you're right. But it's Kristina Halvorson who has applied it to websites and other digital media.

But does it really take an author and consultant to tell us this, when all we need to do is pay more attention to our own behavior when it comes to media? It's truly a classic case of being reminded of something that should already be all too obvious.

Just think about the things we do every day that involve smaller amounts of content, versus things with longer formats.

Most days, we will all read text messages and email, check into a social network, browse some websites, maybe read part of a newspaper and, if we have any free time, perhaps watch a television show or two that is generally a half-hour or hour long.

Things we do *not* do every day: read an entire book (or even part of one), see a movie, watch a miniseries on television, go to the theater.

While some of these choices involve issues of cost, a lot of it just boils down to time and attention. Watching an entire film simply takes a lot of time. So it can't happen every day, and it needs to look pretty compelling for us to schedule it into our busy lives.

Thus, a good deal of our media consumption (more) is made up of smaller amounts of content (less). Less is indeed more.

Halvorson affirms this when it regards websites, as I do when it concerns apps:

- Less content is more user-friendly: As with a well-designed website, an app better deliver its content efficiently and concisely because of our media habits. If we're confronted with something that looks like it will take a lot of our time, people will opt out of it pretty quickly.

- Less content is easier to manage: Let's say that an important feature of your app is its ability to run offline so that it doesn't require an Internet connection to work. This means that all the content is contained within it. What is the schedule for reviewing and updating the project's content? Have you and the content manager agreed on this schedule?

- Less content costs less to maintain: There is a cost to maintaining your content. Somewhere. Perhaps it's paying an editor, or the time it takes for the owner/client/writer to write and review. Or it's the time that you will be billing your client for updating the app with their content changes. In any of these cases, creating a large amount of content up front isn't just a one-time issue: It's an ongoing issue.

I'm not here to scare you away from having a content-rich app. Certainly, some reference applications' value is entirely in their scope and depth. Content has value, so more of it can make an app more valuable. Just be sure that everyone is on board with the responsibilities of properly feeding and watering content over the long term. Consider creating an editorial calendar to help manage these issues.

Maintenance process

In addition to size and schedule issues, the maintenance of app content can vary due to a variety of factors.

iTunes Connect lacks a nifty content management module, so if your content resides inside the app (rather than online), your content management system is Xcode plus your editor of choice—on your laptop or desktop Mac. Taking the example of some reference applications that I collaborate on with a content provider, we always work backwards from my collaborator's desired date for issuing an app update.

We do this because, for the most part, my collaborator is a publisher, the App Store is his bookstore, and I turn the crank of the printing press. Er, I mean Xcode.

So if my collaborator wants an app update to be available around June 1, I need to budget in some time for Apple to review the update (because they review updates just as they review new apps, though the update reviews seem to go a bit faster). Because I have experienced up to a two-week review period, that's what I go with.

Then I add however much time I need to make content updates in the HTML, new images, and so on.

Then I add on a bit more time as a cushion for surprises.

Then I round up a bit, just to be safer still.

And after all of that, I emphasize to my collaborator that getting his changes to me early might help me submit the update early (this rarely works, but it's nice to try).

The maintenance process for content delivered online may be simpler or more complex than this—it really depends on roles and responsibilities. For instance, in the example of the Twitter feed populating an app screen, who maintains the Twitter account? How often? Are the content parameters very refined or more laid-back and casual?

The strength of maintaining content in tiny bites via a ubiquitous tool like Twitter is that anyone can do it nearly anywhere. That's the good thing.

But most people have learned that even with social media, you still need a plan and a process for making it as effective as possible. If you're setting an expectation for fresh content at a particular interval, you better make sure to deliver on time. And if the "you" is a team of people, is everyone in agreement about who the responsible person is?

Finally, remember that content updates in an app might also lead to the updates of App Store screenshots or updates to the app's companion website. Does your maintenance plan cover all of those bases, too?

It should.

Be focused and empathetic, but not alarmed

A colleague of Halvorson's, Erin Anderson, helps point the way to a balanced approach to content management. Applied to managing iOS app content, the approach is to be focused and empathetic, but not alarmed.

Clever. But what does it really mean?

As Anderson reminds us, "Your content can't please all people, all of the time."[2] The trouble with being sensitive, customer-focused designers and content providers is that we are tempted to make every customer or user happy, all the time.

But take that approach to its logical conclusion and it quickly becomes illogical, especially if we pay too much attention to the ratings and comments in the App Store. (Remember how customers can rave—and rant—in that public space?)

Clearly, when we read constructive criticism in an app review (or receive it via email), we should pay attention and act. But being disciplined about following a focused content strategy means that we are really trying to keep most people happy, most of the time. Once we think we can tweak what we deliver to keep everyone happy, we're not only fooling ourselves, we're probably straying from our original plan.

So with regard to making a content strategy, be of good cheer. Realize that just as on the web, the content of your app is everything. That makes it incredibly important, and thus a critical aspect to plan in detail. It's why people will want and use your app. If you insist on keeping your iOS app content-focused, well-maintained, and governed by a plan with a structure and a schedule, you will indeed keep most people happy, most of the time.

And that's called success.

App planning

You know the type: the client or employer who really has no idea what it takes to design their website. This is how they think:

1. Websites run on computers.

2. Computers do things really fast.

3. Web designers work on computers.

4. Therefore, web designers can design websites really fast.

And don't get me started on the clients who think that because I love to design websites, I also love to design their PowerPoint presentation (and they need it in one hour). I mean, they're practically the same thing, right? (And yes, it's usually the same clients.)

But can we really fault our clients for such misunderstandings? Not really. Particularly when we're able to make some of their content or page template updates for them quickly, or they themselves are able to swiftly and easily make site updates because we set them up on some easy-to-use content management system.

Well, the same problem can apply to designing apps, too. They also run on fast little devices, so why wouldn't they also be designed in an instant? OMG—yet another reason for us to simultaneously love and hate those for whom we design things!

Yet this chasm between client perceptions and reality really isn't their fault.

It's our fault.

And we can fix it with some help from Patrick Lynch and Sarah Horton.

Lynch and Horton have been helping people understand the process of designing websites for over 17 years, since they first published the Yale Web Style Guide website in 1993. Now a book in its third edition, *Web Style Guide*[3] is still one of my all-time favorite design books, and here's why: It helps fix client and employee misconceptions about designing websites.

It can also help you design better apps. Read on.

Clarify roles and responsibilities

Most problems with technical projects are not actually technical in nature. They are human. But I don't mean that the problem is the person, though I've thought this and I bet you have, too. Yes, you have—you can admit it.

A lot of these problems can be collected under the broad umbrella of roles and responsibilities. In fact, this perspective helps reinforce that they are people problems, doesn't it? I mean, computers and software don't have roles and responsibilities: They just do exactly what we tell them to, and when we mess up with our directions, they don't do it right either. Because technology doesn't understand roles and responsibilities: It doesn't have any.

So as you start getting excited about a new app project, whether it's for a client or an employer, cool your jets, step back, and start thinking about the various roles that will likely be involved:

- project manager
- information architect
- art director
- editor
- graphic designer
- coder/programmer

And there could be others.

Some of these roles may reside in the same person. Someone will manage the project, someone will structure the content, someone will determine the look and feel, and so on. These roles may involve six people, or two people, or more than six people. It just depends on the size of your project and team.

And in a way, it doesn't really matter how many people are involved (though I generally think the fewer, the better). What matters is that the people who are involved know what they are supposed to do. But how will they know?

You need to tell them.

If you're the only person who knows all the roles, it is your responsibility to define them for your project and have everyone agree about who does what. It's not only in your own best interest to do so (you are setting yourself up to fail if you don't), it's also in the best interest of your app project.

So have a meeting. Write things down. Design a cute team roster document if you need to. Heck, give everyone a team jersey if that's what it takes. But do not proceed with a project until the roles and responsibilities are clearly established.

Develop a project charter

Lynch and Horton offer especially helpful advice when it comes to developing a project charter for web projects and, here again, I find their ideas just as useful for designing app projects.

A project charter is a simple yet comprehensive document that defines the basic parameters of the project. It should ideally begin with the following items:

- the client's (or client's or employer's organization's) mission
- the top two or three goals for the app
- the audience of the app
- a description of how designing this app will meet the app's goals, and support the client's mission
- any measures of success for the app

The more of these bullet points you can define with clarity and brevity (because no one likes to read complicated, nebulous, or wordy goals), the higher the odds that you, your client, and your users will be happy with the app. Similarly, the fewer you define with clarity and brevity, the more ambiguous the project will be. This will lead to false expectations, confusion, misunderstanding, and hard feelings. And once people are unhappy, it makes for a very long and possibly unsuccessful project.

But a project charter need not stop with organizational and programmatic goals. It should also summarize the roles and responsibilities of the people involved as discussed in the previous section. And as long as it's covering these practical matters, how about costs and budgets, deadlines and other milestones along the way, any expected advertising or marketing costs, communications plans, and so forth? Some of these items can eventually become detailed documentation in their own right, but as items in a project charter they just need to be there in some capacity. Make them short-and-sweet list items if possible, and use ballpark estimates when precise figures are not available; this will help you refine them later.

In the end, a project charter should be just a page or two of bullet points, lists, or short sentences. It should be brief enough so that you can review it regularly with the rest of your team—even if your team is just you and your employer or client. And if it is just two people, don't think this excuses you from writing up a project charter. It doesn't: Take any two people without a clear plan, and they are bound to disagree at some point.

Let's face it: A project charter will not guarantee that disagreements won't happen, either. But it will dramatically increase the odds that they are minor rather than major, and thus not showstoppers that sneak up and ruin everyone's excitement for the project.

Diagram your app (or, build it on paper first)

The next step for a good app project plan is to diagram the entire application, screen by screen. This is another step that requires paper. Indeed, designing new media projects is not just using pixels, is it?

So why is a diagram important, if it's faster to just start building? After all, it's all digital—it's pretty easy to rework and make changes.

Sure it is. At first.

Then after the tenth change, you start thinking otherwise and lose sight of the project goal. Scope creep has set in—as your passion for the project quietly creeps away, never to be seen again.

This is your life without planning diagrams.

Planning diagrams don't need to be very sophisticated (though they can be). My suggestion is to start simple. Draw the boxes and label them by hand if you have to—we get so accustomed to using software, we sometimes forget that paper and pencil are still acceptable. And even preferable.

Or, use a program to make a project diagram. Regardless of which tool you prefer, use it to create an easy-to-understand app structure or information architecture (it's just like a structure or information architecture for a website). Show how users of the app will start using it, and where the decision points will lead them. Be clear about how your structure supports the information hierarchies and chunking of information that will be required to make this app elegant and easy to use.

After you have everyone sign off on the diagram, you're one step closer to having your app project not drive you insane.

Save the visuals for the end

It's a bit more difficult to do this with apps than with websites, but try to save most of the visual development and graphic design for the end of the project.

You've probably experienced what it's like to not do this. Clients love to see graphics for a project. Well-designed graphics that glow seductively on a screen make clients (and, let's face it, us) feel great. And they help make it appear that the project is looking complete, even if it isn't.

Remember that false expectation problem? Premature graphics are a leading cause of false project expectations.

But as I already noted, separating the look and feel from the navigation and content of an app can be a bit more difficult. On very small screens, there is less space for the eye candy. Even mocking up a table view navigation can make it look done already—there's just not much to it.

In some cases, you may want to defer building some of the elements that can be quick to build for as long as possible. Remember that simple pencil sketches can get everyone to commit to a relatively detailed design before pixels are ever put to screen.

On the other hand, don't get carried away with excessive documentation or postpone visual elements too long. Over time and with additional experience, you will gain the ability to gauge how much opportunity you need to create for yourself via adequate planning and documentation. Just enough will help answer questions, set realistic expectations, clarify goals, and keep everyone involved happy. Too little can leave gaps for small misunderstandings that can fester into major problems. And too much can drain goodwill from the relationships and energy from the project's momentum.

Don't make planning too big of a stick. Try to make it just right. If you do, happiness and success with your app project is likely to follow.

App usability

Usability consultant Steve Krug, author of *Don't Make Me Think*,[4] has a great set of foundational guidelines in his book about web usability entitled *Billboard Design 101*. These recommendations are based on years of observing how people really use websites: When users visit websites on the large screens of desktop and laptop computers, they don't usually visit in the way that site owners and designers imagine they will.

Site owners who want to share a lot of text content seem to expect that visitors will regard the site as classical literature and take the time to carefully and lovingly read every word. In fact, given the amount of content some site owners are hoping people will read, one might think they're actually frustrated novelists who should seek a publisher instead of putting it all on a website.

As web designers, we have to (perhaps reluctantly) admit that our expectations are similarly high, if not higher. Designers imagine visitors to our gorgeous websites—they are gorgeous, right?—being utterly captivated by the elegant beauty of our designs. They hear choirs of angels singing and feel their eyes well up with tears as they gaze upon our amazing pixel creations in total rapture.

Or something like that.

But Krug, and anyone who takes the time to do usability testing, usually encounters a vast chasm between these utopian expectations and real user behavior. People don't clear their schedules to spend hours visiting and rhapsodizing about websites. Visitors are in a hurry, looking for information, and we can't do much about that. Everything about the online environment speaks to instant gratification, so people behave accordingly: They glance at pages, scan text, and click on things that look like links in order to keep moving.

Take a look at those verbs again: glance, scan, click. They are all measured in fractions of a second. So these are the fractions of seconds we have to win our visitors over with some kind of ode to classical literature or art? Kind of hilarious, when you think about it (no pun intended). Though perhaps Krug should have titled his book *Don't Make Me Laugh* because I'm sure, given his sense of humor, that he laughs hard and laughs often when he observes crestfallen site owners and designers sit in on usability evaluations. I've been there, and I know exactly what it feels like to have my high design expectations lowered.

So what does Krug recommend for this situation, and how does it apply to designing apps for iOS mobile devices? Let's take a look, and think about it, so our app users don't need to.

Create a clear visual hierarchy

Establishing a clear visual hierarchy is one of the most important, central services that designers can bring to bear on someone's project and its content. In fact, it's timeless: I was teaching undergraduates at the University of Minnesota the fundamentals of visual hierarchy in Graphic Design 101 way before they were learning much about sexy web and interactive media technologies, because it's just as important for designing an effective poster, book, or postcard as it is for designing a good website and, now, mobile application.

Krug argues that the basic tenets of good visual hierarchy are

- Make the things that are most important the most prominent.
- Group items that belong together (and separate those that do not).
- Use nesting to add further clarity about relationships between categories and items within them.

I wholeheartedly agree with these basic rules. They have governed the printed page for centuries and allowed people to navigate large amounts of dense content (think about newspapers and dictionaries).

Some iOS suggestions regarding these guidelines:

- Be just as savvy with typography in apps as you would be with websites, but favor even fewer styles. Sizing, the use of bold, and at most a second typeface should take you as far as you need to go in creating content hierarchy on a small screen.
- Use grouped table views when you have groups of items that belong together.
- Definition lists are a nice way to nest information under labels.

Take advantage of conventions

Conventions are a bit of a paradox for designers. We like great ideas, yet the more that great ideas become imitated by others and broadly adopted into the mainstream (thus becoming a convention), the more we start to chafe a bit and want to do something different. Just to be unique and creative. To design.

But as Krug reminds us, conventions are your friends. And designing is not always creating something different.

The more that conventions are adopted, the less users and customers have to figure things out on the fly as they try to use what we have designed. And if getting people to information quickly and easily is our goal, conventions should play a big role. If instead we want an adventurous, challenging means to uncovering a tidbit of information (e.g., an "Easter egg"), then designing an obscure path is the way to go.

The latter might be great fun in the realm of gaming, but the fact is that most people want quick results. This usually means using more familiar items than unfamiliar items.

Consequently, this book emphasized becoming familiar with iOS conventions like the status bar, title bar, table view, and tab navigation for an important reason: We see these items all the time in other mobile applications (including the ones that Apple designs), so we need to learn how to design apps that fit the same mold when it's appropriate. And it's appropriate every time we want to design content-based apps that convey information easily to users whenever, and wherever, they seek it.

Break pages up into clearly defined areas

Given how small iPod touch and iPhone screens are, you might at first dismiss having to think about defined areas in mobile applications.

Think again.

Granted, the smaller iOS screen size doesn't really lend itself to much dead space like a web page can. So the problem of drawing attention to important content on a screen is not necessarily as challenging in the small screen environment as in the large screen environment.

Nonetheless, as soon as you begin laying out an app screen with a title bar, content area, and perhaps a tab navigation, you're making decisions about actionable areas versus content to read. And despite the small screen, it's still possible to place additional links throughout an app screen in such a way that there's too much for a user to sort through. Similarly, structuring your content in the wrong way—or mislabeling navigation buttons or tabs—reduces the clarity of your intended design.

As soon as you enter iPad territory, issues like clearly defined areas of a page can become much trickier (**Figure A.1**).

A.1 The App Store on the iPad has a lot going on. Are the areas clearly defined, and hierarchies well established? Is this screen designed to get visitors somewhere specific or to encourage them to browse?

Of course, there are exceptions to some of these rules—but rarely, if ever, all of these rules. Looking at the example of the App Store, it's clear to me that Apple's design team can do a good job of clearly defining areas of a screen: There are three tabbed choices at the top, an In the Spotlight coverflow area beneath that, and areas below that focus on some of the newest apps that are available. The screen layout is quite clear, and it uses some emerging conventions of the device quite well—but it's fair to say that the hierarchy isn't particularly clear.

But when we think about it, Apple is trying to get people interested in lots of apps. All of the apps. So the lack of hierarchy is probably intentional: This is an app design that encourages browsing and searching, not navigating structured content.

So feel free to stray from the good usability conventions—when you can justify it.

Make it obvious what's clickable

People do a lot of clicking on both websites and iOS apps, so making it obvious what is clickable—or, in this case, touchable—is of paramount importance for ease of use and efficiency. Fortunately, Apple provides us with several UI standards for designing obvious touchable spots in our apps:

- rows in table view navigations
- tabs in tab bar navigations
- rectangular buttons with rounded corners
- back buttons that point left

Additionally, more advanced app designers can leverage the nice date picker interface and other ways to interact with information and choices.

So the basic tenet of touchability is this: If something links to something else in an app, it better be obvious. I still find myself designing hyperlinks within app content and later realizing that sometimes they're just not obvious enough as links. On the other hand, peppering content with a bunch of buttons breaks up the flow of reading, too (**Figure A.2**).

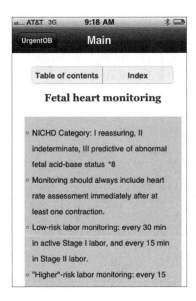

A.2 A screenshot from one of the apps I designed. In my opinion, there's a button or link that should be made more obvious. Can you tell which one it is? It's the *8, which is bold but otherwise difficult to pick out as a link.

Hmm, designing easy-to-use app content is not always simple, is it? In light of this, never hold back on iterating to keep improving a design and applying new ideas that you learn from app to app. You can update apps once they're in the App Store: Take advantage of this!

Minimize noise

Unnecessary visual noise is less likely to infiltrate an iOS app screen than a much larger web page, but it's not impossible. Often noise is in the details, such as the darkness of lines or shadows. If you're designing a custom button, how wide can the border be—and how dark the shadow effect—before the button begins to make too much noise? It's the same with a custom table view: If you're adding a textured background image, what is elegant and subtle, and what is just loud and heavy-handed?

And once again with the iPad screen: The more real estate you have to work with, the more opportunity you have to mess it up.

In my opinion, UI design for the iPad is still very much a Wild West scenario. It's a lot like the web in the mid-'90s: Nearly anything goes, simply because it takes a while for conventions to emerge, and tablet computing is a very new area to work in.

But don't fall into the trap of thinking that because the conventions are still emerging, you can design whatever you want. Keep these usability recommendations in mind even more when designing for the iPad. Try to come up with good, elegant, and usable ideas that feel familiar.

Perhaps you can help develop the best practices that end up being everyone's design conventions.

REFERENCES

1. *Content Strategy for the Web*, Kristina Halvorson (New Riders, 2010)

2. http://blog.braintraffic.com/2010/08/content-creation-quality -vs-quantity-or-"a-recipe-for-content-deliciousness"/

3. Yale University Press, 2008. And it's still published as a website, too: http://www.webstyleguide.com.

4. *Don't Make Me Think*, Steve Krug (New Riders, 2006).

INDEX

WATCH
READ
CREATE

Meet Creative Edge.

A new resource of unlimited books, videos and tutorials for creatives from the world's leading experts.

Creative Edge is your one stop for inspiration, answers to technical questions and ways to stay at the top of your game so you can focus on what you do best—being creative.

All for only $24.99 per month for access—any day any time you need it.

creative
edge

peachpit.com/creativeedge